10.50

Nonlinear System Theory

A Functional Analysis Approach

PRENTICE-HALL NETWORKS SERIES

Robert W. Newcomb, *editor*

Nonlinear System Theory

A Functional Analysis Approach

JACK M. HOLTZMAN

Bell Telephone Laboratories
Whippany, New Jersey

Prentice-Hall, Inc., Englewood Cliffs, New Jersey

To Natalie

Current printing (last digit):

10 9 8 7 6 5 4 3 2 1

13-623264-7
Library of Congress Catalog Card No. 76-132082
Printed in the United States of America

PRENTICE-HALL INTERNATIONAL, INC., *London*
PRENTICE-HALL OF AUSTRALIA, PTY., LTD., *Sydney*
PRENTICE-HALL OF CANADA, LTD., *Toronto*
PRENTICE-HALL OF INDIA PRIVATE LIMITED, *New Delhi*
PRENTICE-HALL OF JAPAN, INC., *Tokyo*

Preface

This book is an exposition on a number of topics in nonlinear system theory using some notions of functional analysis. The functional analysis viewpoint provides a broad and powerful means for treating many problems in nonlinear systems as well as in other fields (e.g., filtering, optimal control). Therefore, learning some functional analysis should prove to be very much worth the effort. This effort can be minimized in this book by recognizing that knowledge of only a small portion of functional analysis can be used to obtain many useful results. It is thus possible to start with fundamentals and then rather quickly lead the reader to some of the frontiers in the field.

Nonlinear system theory is so broad that no book can provide a fully comprehensive discussion on this subject. The topics are selected to cut across a number of interesting areas. An idea of the scope of the book can be gleaned by inspection of the Contents and by reading the Introduction. Incidentally, the title is not meant to imply that I am presenting *the* theory of nonlinear systems by which all nonlinear problems can be treated. Rather, my intent is to present some theory which I believe to be useful in many aspects of nonlinear systems.

I hope that this book will be useful to engineers and also, be of interest to some mathematicians, especially in directing their attention to unsolved problems. Although the book is more mathematical than the usual engineering book, I try to strike a balance between what is proved and what is stated to be provable. To give proofs of everything would amount to spoonfeeding in the case of the simpler proofs and digressing too far in the case of some of the more difficult proofs. The book is mainly concerned with mathematical theorems as tools for nonlinear analysis and I try to resist the temptation of dwelling on the intrinsic interest of the theorems themselves.

v

The prerequisites for reading the book are somewhat dependent on the intangible maturity of the reader. Some familiarity with matrices, differential equations, and linear systems is needed. Advanced calculus is highly desirable. Some complex variable theory is used in Chapter VIII although mostly in an appendix. The book can be used by engineers as a second exposure to nonlinear systems (the first course typically covers such topics as phase-plane, describing functions, and Liapunov stability). The text has been used in the second half of a year graduate engineering course in nonlinear systems. The book might also be used in a course for applied mathematicians to show some applications of functional analysis.

I have tried to consistently use notation which is also consistent with that generally used in reference material. However, this approach has a limitation since much of the terminology is not standardized. Two conventions should be mentioned in particular. First, a nonlinear operator is an operator not necessarily linear (not an operator which is not linear). This results in an economy of words in several places. Second, we do not always keep to the convention that x or $x(\cdot)$ is a function with $x(t)$ the value of the function x at t. We keep to the convention in early chapters where this distinction must be learned if not already known beforehand. But in later chapters, it becomes inconvenient especially in connection with transforms (e.g., the Fourier transform of $f(t)e^{-\sigma t}$).

It is a pleasure to acknowledge the suggestions and comments I've received, particularly from R. S. Ellis, L. J. Forys, H. Heffes, S. Horing, S. H. Kyong, and R. P. Marzec. Mrs. Mary Lou Martinez efficiently handled the typing and Mrs. Sharon Miller, the programming. The help of Mrs. Julie Picariello is also appreciated. I am grateful to Bell Telephone Laboratories for its consistent support. My wife Natalie was always a source of inspiration and her help with the proofreading was invaluable.

J. M. HOLTZMAN

Whippany, New Jersey

Contents

Introduction

In order to appreciate the phenomena associated with nonlinear systems, it is best first to review the relative simplicity associated with linear systems. Let us consider a linear system to be a relationship between an input x and an output y symbolically denoted by $y = Lx$ which satisfies

$$L(a_1 x_1 + a_2 x_2) = a_1 L(x_1) + a_2 L(x_2).$$

We immediately observe two very attractive features of a linear system. First of all, the sum of two inputs results in the sum of the responses to the individual inputs (principle of superposition). Secondly, a multiple of an input results in a multiple of the output. This leads to the possibility of obtaining knowledge about all the solutions from information on only some solutions.

With nonlinear systems, we cannot count on the above nice properties. With two different inputs, a nonlinear system can exhibit two drastically different outputs. Because of the difficulties involved in nonlinear analysis, approximation methods are commonly used. In fact, one often tries to adapt the relatively simple methods for linear systems to nonlinear problems. This must be done, of course, with some hedging. Indeed, some nonlinear behavior is completely unexplainable by linear methods. We now give two typical examples of approximate methods.

Duffing's equation

$$\ddot{y} + ay + by^3 = f \cos \omega t \qquad a > 0$$

is a much studied equation because of the rich diversity of its behavior. With $b = 0$, the equation can clearly represent a periodically forced linear system consisting of a mass and a spring (or an electric circuit with inductance and capacitance). With $b \neq 0$, the spring is nonlinear. If the stiffness s is defined as the derivative of the restoring force, then $s = a + 3by^2$. If $b > 0$, the stiffness increases with y; this is called a hard spring. If $b < 0$, we have a soft spring. Duffing's equation could also represent an approximation to the pendulum equation,

$$\ddot{y} + \frac{g}{l} \sin y = 0 \qquad (g = \text{gravitational constant}; \, l = \text{length})$$

by taking the first two terms of the series expansion of $\sin y$.

Given a nonlinear equation like Duffing's equation, the question immediately arises as to how to solve it. Most nonlinear differential equations cannot be solved exactly by paper and pencil analysis. Computers can provide numerical solutions for given parameters and initial conditions. In this book, we are more interested in the general behavior of the solutions rather than in particular solutions.

Let us discuss the nature of solutions to Duffiing's equation. Since it is not exactly solvable except for very special selection of the parameters, we are led to try to obtain an approximate solution. Since the forcing function is periodic, one suspects that the solution is also periodic. Hence as a first cut, try the simplest periodic function

$$y(t) = A \cos \omega t$$

as a solution. Putting this into Duffing's equation yields

$$-A\omega^2 \cos \omega t + aA \cos \omega t + \frac{bA^3}{4}(3 \cos \omega t + \cos 3 \omega t) = f \cos \omega t.$$

We see that we cannot generally balance the trigonometric terms on both sides of the equation unless $b = 0$. An approximation is obtained if we just balance the coefficients of the $\cos \omega t$ terms and ignore the $\cos 3\omega t$ term. This is certainly plausible if b is "small." This method of approximation is called harmonic balance. With Duffing's equation, it leads to the following equation:

$$\tfrac{3}{4}bA^3 + (a - \omega^2)A - f = 0.$$

We shall see that for given a, b, f, ω there may be more than one real A

satisfying this equation. That is, for the same forcing function ($f \cos \omega t$) the method of harmonic balance may indicate that there may be more than one type of solution. This kind of behavior is called jump resonance; a small change can result in a jump from one type of solution to another.

Let us note another type of peculiar behavior of nonlinear systems which again can occur even with an equation as simple and innocent looking as Duffing's equation. If instead of the approximation $y(t) = A \cos \omega t$, we try $y(t) = A_{1/3} \cos \frac{1}{3} \omega t + A_1 \cos \omega t$ and balance the coefficients of the $\cos \frac{1}{3} \omega t$ and $\cos \omega t$ terms, we find that the existence of subharmonics may be indicated. By subharmonic, we mean that the solution of the differential equation has a lower fundamental frequency than the forcing function.

Hence, using a rather simple approximation technique, we can quickly gain some insight into nonlinear phenomena. However, we should like to have some confidence that the approximate solutions are reliable indications of the true solutions. That is, is there an exact solution near the approximation? This question is not raised just for esoteric or academic reasons. For example, the existence of subharmonics is not something that can be obviously established from physical reasoning. In fact, one might argue the other way: Physical intuition could be built up by knowledge of such nonlinear phenomena as subharmonics whose existence are established experimentally or mathematically (this book concentrates on mathematical methods).

The above considerations lead us to one orientation of the book. We consider approximate techniques along with their justification. It must also be recognized, however, that the use of approximate methods (with apparent "customer satisfaction") often precedes their justification. We thus wish to clearly point out not only how much justification is available for an approximation technique but also what are the known limitations of the method of justification. This delineates areas for further study.

Let us consider a simplified version of another type of approximation technique to be considered. Let $x_0(t)$ be a solution of the scalar differential equation

$$\dot{x}(t) = f(x(t), p)$$

when the parameter p takes on the value p_0. For p close to p_0, we can approximate the change in x in the following manner:

$$\dot{x}(t) - \dot{x}_0(t) = f(x(t), p) - f(x_0(t), p_0)$$

$$= \frac{\partial f}{\partial x}(x_0(t), p_0)(x(t) - x_0(t)) + \frac{\partial f}{\partial p}(x_0(t), p_0)(p - p_0) + \cdots$$

$$\delta \dot{x}(t) = \frac{\partial f}{\partial x}(x_0(t), p_0)\delta x(t) + \frac{\partial f}{\partial p}(x_0(t), p_0)\delta p.$$

In the above, we have used a series expansion (assuming it to be justified) and the three dots represent higher order terms. By neglecting the higher order terms, we obtain a differential equation for $\delta x \cong x - x_0$. The differential equation in δx is linear though generally time-varying; it is the linearity of the equation which makes it appealing to work with. This technique of linearizing about a nonlinear solution is widely used. We shall put it into a more general setting and indicate how to determine the accuracy of negelecting the terms represented by the three dots.

The problems just discussed are concerned with differential equations, perhaps the most common representation of nonlinear systems. Some nonlinear systems are naturally described in terms of a nonlinear element and a linear element. In many cases, it is possible and convenient to represent the linear element by means of a transform (Laplace or Fourier). With linear systems, transform representations are familiar and lead to results which are elegant in their simplicity (notably Nyquist's stability criterion). The use of transforms is also compatible with how linear elements can be readily measured (frequency response determination) and modified (e.g., the use of compensation networks in linear servo theory). Hence, we shall put some emphasis on using transforms.

With the use of Fourier transforms, we shall get neat stability conditions for nonlinear feedback systems. The conditions are strongly reminiscent of Nyquist's criterion. The problem of obtaining conditions for stability (which we shall precisely define) is clearly in the spirit of obtaining general information on the solutions rather than trying to obtain particular solutions.

We have briefly discussed three types of nonlinear problems, all not apparently closely related. They can, however, be discussed in a common framework using certain notions of mathematics, particularly from the subject of functional analysis. Hence our first objective is to supplement the mathematical background of those readers not familiar with the required results.

I

Mathematical Background

This chapter is a brief introduction to some mathematical notions that we shall find useful throughout the book. In particular, some aspects of functional analysis will be discussed. An introduction to functional analysis could easily take as many pages as are contained in the present book. Hence, the scope of this chapter is limited to stating some needed results (without proving many of them); only a small portion of the results of functional analysis are needed for our purposes.

In later chapters, additional mathematical background will be introduced where required. The functional analysis novice will thereby be enriching his functional analysis background while proceeding with the book's analysis of nonlinear problems. Thus, this chapter is intended to be a "starter" for the functional analysis novice (who should try to do all the exercises and consult the references). For those familiar with functional analysis, this chapter serves to fix the definitions and notation and to make required results available for handy reference.

I.1 SETS

A set is, generally speaking, a collection (class, family, aggregate) of objects (members, elements, points). This very general notion of set gives the important connotation of membership. If X is a set and x is a member of X, we write $x \in X$. That x is not a member of X is denoted by $x \notin X$. If all the members of one set X are also members of another set Y, then X is said to be a subset of Y and we write $X \subset Y$. The two sets X and Y are equal, $X = Y$, if they have the same members; this is equivalent to $X \subset Y$ and $Y \subset X$.

The set $\varnothing$ with no members is called the void set, the null set, or the empty set. If X is a set, then $\varnothing \subset X$ since there are no members of $\varnothing$ not in X.

The union of two sets X and Y, denoted by $X \cup Y$, is defined as the set of all points which belong to at least one of the sets. The intersection of X and Y, $X \cap Y$, is the set of all points belonging to both X and Y (see Figure I.1).

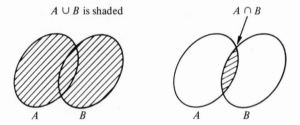

Figure I.1 Union and intersection

Specific sets are usually defined and described with the following notation: $\{x : \text{proposition about } x\}$. This bracket notation means the set of all x such that the proposition about x is true. Note that x is a dummy variable and may be replaced by any other variable not used in the proposition.

If $Y \subset X$, then the set $\{x : x \in X, x \notin Y\}$ is a subset of X and is called the *complement* of Y with respect to X (or the complement of Y if there is no ambiguity as to what X is).

An important set is the real line which is the set of all real numbers,

$$R = \{x : x \text{ is a real number}\}.$$

Intervals of the real line are denoted as

$$[a, b] = \{x : x \in R, a \le x \le b\},$$
$$(a, b) = \{x : x \in R, a < x < b\},$$
$$[a, b) = \{x : x \in R, a \le x < b\},$$
$$(a, b] = \{x : x \in R, a < x \le b\}.$$

With this notation, $R = (-\infty, \infty)$. While we are considering R, it is useful to recall the definitions of supremum (least upper bound) and infimum (greatest lower bound). Suppose A is a nonempty subset of R and for all $x \in A$, $x \leq M$. If $M \in A$, then M is the maximum of the set A. Some sets, e.g., $A = [0, 1)$, have no maximum. Clearly, any $M_1 > M$ will also satisfy $x \leq M_1$. The least M satisfying $x \leq M$ (which always exists) is called the supremum of A and denoted by sup A. If $A = [0, 1)$, then sup $A = 1$ but, as previously mentioned, the maximum does not exist. If $A = [0, 1]$ then sup $A = \max A = 1$. The infinum (greatest lower bound) is analogously defined. If $A = (0, 1]$, inf $A = 0$ but min A does not exist. If $A = [0, 1]$, inf $A = \min A = 0$.

The *product* of two sets X and Y, denoted $X \times Y$, is the set $\{(x, y) : x \in X, y \in Y\}$ where (x, y) is an *ordered pair* which has the property $(x_1, y_1) = (x_2, y_2)$ if and only if $x_1 = x_2, y_1 = y_2$. The plane, R^2, is the product set $R \times R$.

It is sometimes necessary to refer to the size of a set. For sets with a finite number N of elements (called a finite set), we merely count and say the size is N. If the set A has an infinite number of members, we denote its size by putting it in one-to-one correspondence with another set B which is perhaps more familiar. By one-to-one correspondence we mean a rule that assigns to each $a \in A$ a unique $b \in B$ and to each $b \in B$ a unique $a \in A$. This type of gauging reduces to counting for finite sets but is more general. A set in one-to-one correspondence with the set of positive integers is called *denumerable*. The set of all integers may be shown to be denumerable. A set is *countable* if it is either denumerable or finite. An *uncountable* set is one which is not countable. The interval $[0, 1]$ is uncountable.

I.2 METRIC SPACES

It is of interest to introduce a notion of distance between points in a set. To do this we define a metric space as a pair of two things: a nonempty set X and a metric d. The metric d is a measure of distance between elements of X and is a non-negative real function satisfying the following three conditions for arbitrary $x, y, z \in X$:

1. $d(x, y) = 0$ if and only if $x = y$
2. $d(x, y) = d(y, x)$ (symmetry)
3. $d(x, z) \leq d(x, y) + d(y, z)$ (triangle inequality)

We list below some metric spaces. As an exercise, the reader should verify that the conditions for a metric are satisfied.

1. The real line R with the usual notion of distance between two points

$$d(x, y) = |x - y|.$$

2. An arbitrary nonempty set with what is sometimes called the trivial metric

$$d(x, y) = \begin{cases} 0 & x = y \\ 1 & x \neq y. \end{cases}$$

This metric is sometimes useful to construct counterexamples to disprove conjectures about metric spaces.

3. The set of all continuous real-valued functions defined on $[a, b]$ with $b - a < \infty$ and

$$d(x, y) = \max_{t \in [a,b]} |x(t) - y(t)|.$$

In the last case, the definition of countinuity is as given in calculus books (and repeated in Sec. I.6). A general definition of continuity for metric spaces will be given in Sec. I.4. The space in the third example is denoted $C(a, b)$. This metric is useful in error analysis where one wishes to know the maximum deviation between an approximation and an exact solution to some equation.

Other metric spaces will be introduced later in the book. Note that from the description of spaces (1) and (2), two different metrics may be associated with R which emphasizes the point that a metric space is a set along with a metric; different metrics result in different metric spaces. However, with a slight abuse of notation, reference to metric spaces is often made by mentioning only the set, it being understood what metric is being used. In particular, if R is considered as a metric space and the metric is not specified, the metric is understood to be $d(x, y) = |x - y|$.

The use of a metric in a set allows a definition of convergence. Intuitively, an infinite sequence $\{x_n\} = \{x_1, x_2, \ldots\}$ converges to a point x if as n gets larger and larger, the distance between x_n and x gets smaller and smaller. Mathematically, $\{x_n\}$ converges to a *limit* x, denoted by $\lim_{n \to \infty} x_n = x$ or $x_n \to x$, if for any $\epsilon > 0$, there is an n_0 such that $n \geq n_0 \Rightarrow d(x, x_n) < \epsilon$. ($\Rightarrow$ means implies.)

Exercise I.2.1

Prove the following:
a. $d(x, y)$ is a continuous function. That is, $x_n \to x$, $y_n \to y \Rightarrow d(x_n, y_n) \to d(x, y)$.
b. A convergent sequence can have only one limit.

If $x \in X$, a metric space, an *open ball* $S_r(x)$ of radius $r > 0$ is defined as the set $\{y : y \in X, d(x, y) < r\}$. The *closed ball* $\overline{S_r(x)}$ of radius $r > 0$ is the set $\{y : y \in X, d(x, y) \le r\}$. The point x is an interior point of $A \subset X$ if there is some open ball $S_r(x)$ which is a subset of A.

If $x \in X = R$, then $S_r(x) = (x - r, x + r)$ and $\overline{S_r(x)} = [x - r, x + r]$.

If $X = C(0, 1)$ and $x_0(t) = \frac{1}{2}t^2$, then $S_{1/4}(x_0)$ is pictorially shown in Figure I.2.

With $A \subset R$, and $A = [0, 1]$, any $x \in (0, 1)$ is an interior point of A but $x = 0$ and $x = 1$ are not. A subset of X is said to be *open* if all of its members are interior points. As subsets of R, the interval $(0, 1)$ is open, but $[0, 1]$ is not (neither is $[0, 1)$ nor $(0, 1]$).

A *neighborhood* of a point x is a set containing an open set which has x as a member.

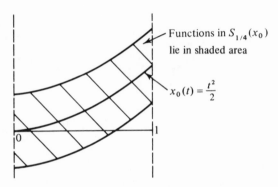

Functions in $S_{1/4}(x_0)$ lie in shaded area

$x_0(t) = \dfrac{t^2}{2}$

Figure I.2 Closed ball in $C(0, 1)$

The point $x \in X$ is said to be a *limit point* of $E \subset X$ if there is a sequence $\{x_n\}$ with $x_n \in E$, $x_n \ne x$, $n = 1, 2, \ldots$, which converges to x.[†] The *closure* of E, denoted by $\bar{E}$, is the set of all points in E and the limit points in E, i.e.,

$$\bar{E} = E \cup \{x : x \in X, x \text{ is a limit point of } E\}.$$

If $E = \bar{E}$, E is said to be *closed*. As subsets of R, the interval $[0, 1]$ is closed but $(0, 1)$ and $(0, 1]$ are not because 0 is a limit point of these two sets which does not belong to the sets.

With the above commonly used notation, there is a slight inconsistency between closed ball $\overline{S_r(x)}$ and closure of a set. It may be shown that the closure of $S_r(x)$ is contained in $\overline{S_r(x)}$ but the following example shows that the closure of $S_r(x)$ need not equal $\overline{S_r(x)}$. Using a metric space X and the trivial metric, $S_1(x) = \{x\}$ and the closure of $S_1(x) = \{x\}$, but $\overline{S_1(x)} = X$. However, it is true that the closure of $S_r(x) = \overline{S_r(x)}$ in normed linear spaces which we introduce

[†] See Porter p. 25 for a clarification of the distinction between limit and limit points.

in the next section. What is important for our purposes is that a closed ball is a closed set in a metric space.

Exercise I.2.2

Prove that the complement of a closed (open) set is open (closed).

Exercise I.2.3

Prove that both the metric space X and the empty set are both open and closed (this shows that closed and open are not the opposites of each other).

The notion of convergence of a sequence of points (as defined above) is most important. To directly use the definition to determine whether a given sequence converges one would have to know the point to which it converges. To get around this, we define a *Cauchy sequence* as a sequence $\{x_n\}$ such that $d(x_n, x_m) \to 0$ as $n, m \to \infty$, i.e., for every $\epsilon > 0$, there is an $N(\epsilon)$ such that $n, m \geq N(\epsilon)$ implies $d(x_n, x_m) < \epsilon$. It is easily shown that a convergent sequence is a Cauchy sequence (exercise for the reader) but it is not true in all metric spaces that a Cauchy sequence is necessarily a convergent sequence. Those metric spaces for which it is true, we call *complete*. Thus, in a complete metric space a necessary and sufficient condition for a sequence to be convergent is that it be Cauchy. Examples of complete spaces are R and $C(a, b)$.

If we have a nonempty closed subset A of a complete metric space X, the set A by itself can be considered to be a metric space. When considered by itself to be a metric space it is complete for if a sequence in A is Cauchy, it converges to an element in X due the completeness of X but then it must converge to an element of A due to the closedness of A.

I.3 NORMED LINEAR SPACES

In Sec. I.2, we added structure to sets by introducing a distance function. We can also add structure to a set by introducing algebraic operations (addition, scalar multiplication). To this end we define a *linear space* as a set A for which the following conditions are satisfied:

    I. For any $x, y \in A$, there is a uniquely defined $z = x + y$ with $z \in A$ and such that

        1. $x + y = y + x$,
        2. $x + (y + w) = (x + y) + w$ any $w \in A$,
        3. there is an element $0 \in A$ such that $x + 0 = x$ for all $x \in A$,

    4. for every $x \in A$, there is an element $-x \in A$ such that $x + (-x)$ $= 0$.

II. For every $\lambda \in \mathscr{F}$ (the set of real or complex numbers) and $x \in A$ there is defined an element λx (the product of x and λ) such that

    5. $\lambda(\mu x) = (\lambda\mu)x \quad\quad \mu \in \mathscr{F}$,

    6. $1 \cdot x = x$,

    7. $(\lambda + \mu)x = \lambda x + \mu x \quad\quad \mu \in \mathscr{F}$,

    8. $\lambda(x + y) = \lambda x + \lambda y \quad\quad$ any $y \in A$.

The linear space is called real or complex depending on whether $\mathscr{F}$ (called the *field*†) is the set of real or complex numbers. It is immediately recognized that the above eight properties are familiar ones for, say, the real line or plane.

A nonempty subset of a linear space A is said to be *convex* if for any two elements $x, y \in A$ all points on the line connecting x and y also belong to A. This is expressed mathematically as: $x, y \in A \Rightarrow \theta x + (1 - \theta)y \in A$ for all $\theta \in [0, 1]$. See Figure I.3 for examples of convex and nonconvex sets in the plane.

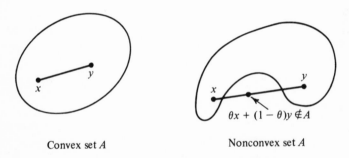

Convex set A Nonconvex set A

Figure I.3 Convexity in the plane

Many important metric spaces are often also linear spaces. The usefulness of the space being both metric and linear is considerably enhanced when the metric is defined in a way that is related to the space being linear. This is done with a norm which we introduce now.

A linear space X is called a *normed linear space* if every element $x \in X$ has associated with it a real nonnegative number called its norm, denoted by $\|x\|$ and satisfying the following conditions:

    (i) $\|x\| = 0 \Leftrightarrow x = 0$

    (ii) $\|\lambda x\| = |\lambda| \cdot \|x\| \quad\quad \lambda \in \mathscr{F}$

    (iii) $\|x + y\| \leq \|x\| + \|y\| \quad\quad$ (triangle inequality)

---

† A complete definition of field is given in Dieudonné, p. 15.

If, for $x, y \in X$, we set

$$d(x, y) = \|x - y\|, \tag{I.3-1}$$

then X becomes also a metric space. The first two conditions for a metric are clearly satisfied. The third condition follows easily from condition (iii) imposed on the norm:

$$d(x, y) = \|x - y\| = \|(x - z) + (z - y)\| \leq \|x - z\| + \|z - y\|$$

$$= d(x, z) + d(z, y). \tag{I.3-2}$$

Two examples of normed linear spaces are the real line R with $\|x\| = |x|$ and the space of real-valued continuous functions $C(a, b)$ with $\|x\| = \max_{t \in [a,b]} |x(t)|$. Note again that there is sometimes a slight lack of consistency when we refer to, say, R as being simultaneously a set, a metric space, and a normed linear space. However, from the context in which it is being used, no confusion should arise. A metric space with the trivial metric cannot be made into a normed linear space with (I.3-1) because condition (ii) in the definition of norm would be violated.

Some simple properties of norms are:

(a) $| \|x\| - \|y\| | \leq \|x - y\|$ $\qquad\qquad\qquad$ (I.3-3)

(b) $\|x\|$ is a continuous function of x, i.e., $x_n \to x_0 \Rightarrow \|x_n\| \to \|x_0\|$

Exercise I.3.1

Prove the above properties of norms.

Since a normed linear space is a metric space if we set $d(x, y) = \|x - y\|$, it may also be a complete space. A complete normed linear space, called a *Banach space*, has the property that a Cauchy sequence $\{x_n\}$ (i.e., $\|x_m - x_n\| \to 0$ as $m, n \to \infty$) converges to an element in the space. The normed linear spaces R and $C(a, b)$ are Banach spaces. When discussing a normed linear space as a metric space we shall always assume that (I.3-1) is satisfied (i.e., the metric is related to the linear structure of the space).

A *subspace* of a linear space is a subset which by itself satisfies all the axioms of a linear space. When a closed subspace of a Banach space is considered to be a normed linear space by itself it is complete (see the last paragraph of Sec. I.2) and thus a Banach space. An example of a subspace is the set of real numbers as a subset of the set of complex numbers (with absolute value for a norm).

I.4 MAPPINGS; LINEAR OPERATORS

Let X and Y be two arbitrary nonempty sets. If $\Omega \subset X$ and if to each $x \in \Omega$ there corresponds one and only one $y \in Y$, denoted by $F(x)$, we say F is a *mapping* (or function or transformation or operator) of Ω *into* Y. Ω is called the *domain* of F and $\{y : y = F(x), x \in \Omega\}$ is called the *range* of R. If the range is Y, then F is a mapping of Ω *onto* Y.

If X and Y are metric spaces, the mapping F is *continuous* at $x_0 \in \Omega$ if $x_n \to x_0 (x_n \in \Omega)$ always implies $F(x_n) \to F(x_0)$. This is equivalent to: for any $\epsilon > 0$, there is $\delta > 0$ such that, if $y \in \Omega$, then $d(x_0, y) < \delta$ implies $d(F(x_0), F(y)) < \epsilon$. Note that $d(x, y)$ uses the metric for X and $d(F(x), F(y))$ uses the metric for Y so that use of the same symbol for two different metrics is a slight, but common, abuse of notation which leads to no confusion. If F is continuous at all $x \in \Omega$, we say that F is continuous on Ω. Note that, given an $\epsilon > 0$, the δ depends in general on ϵ and x. If it does not depend on x, then F is *uniformly continuous*.

Another note concerning notation. For $F(x)$ we sometimes write Fx.

We shall have occasion to deal with one function succeeding another. Let f map X into Y and g map Y into Z. Then $f(x) \in Y$ and $g(f(x)) \in Z$ or gf maps X into Z. The mapping gf is called the *composition* of g and f. We can easily extend the definition of composition to more than two functions, e.g., fgh, or $fff = f^3$, etc. The distinction between composition and multiplication of two functions will be either obvious or specified.

Exercise I.4.1

Show that gf is continuous if both f and g are.

A simple but important mapping is the *identity operator* of X onto itself, denoted by I. If $x \in X$, then $Ix = x$. A simple mapping of a linear space into itself is the *null operator* which maps any element into 0, the zero element.

Let X and Y be two linear spaces (both real or both complex). An operator A mapping X into Y is said to be *linear* if

$$A(\alpha_1 x_1 + \alpha_2 x_2) = \alpha_1 A x_1 + \alpha_2 A x_2 \qquad (\text{I.4-1})$$

for all $x_1, x_2 \in X$ and for all $\alpha_1, \alpha_2 \in \mathscr{F}$. If Y is the set of real or complex numbers, then A is a *linear functional*. If X and Y are normed linear spaces, the operator A is said to be *bounded* if there is some $M < \infty$ such that

$$\|Ax\| \leq M\|x\| \qquad (\text{I.4-2})$$

for all $x \in X$ (again note that there is a slight lack of precision of notation in using the same symbols for the norms of Ax and x but no ambiguity arises).

Theorem I.4.1

If A is a linear transformation of the normed linear space X into the normed linear space Y, then continuity of A is equivalent to its boundedness.

Proof. First we show that boundedness implies continuity. Letting $x, y \in X$, we have $\|Ax - Ay\| = \|A(x - y)\| \leq M\|x - y\| \leq M\delta$. Then, referring back to the definition of continuity, we take $\delta < \epsilon/M$ and A is continuous.

To show that continuity implies boundedness, assume that A is not bounded; this will be shown to lead to a contradiction. For A not to be bounded there must exist a sequence $\{x_1, x_2, \ldots\}$ in X such that $\|Ax_n\| > n\|x_n\|$. Setting

$$z_n = \frac{x_n}{n\|x_n\|}, \tag{I.4-3}$$

we see that $\|z_n\| = 1/n$, i.e., $z_n \to 0$ as $n \to \infty$. Now consider the sequence $\{Az_1, Az_2, \ldots\}$. For this sequence,

$$\|Az_n\| = \frac{\|Ax_n\|}{n\|x_n\|} > \frac{n\|x_n\|}{n\|x_n\|} = 1. \tag{I.4-4}$$

But $A(0)$ must be zero for a linear operator and $z_n \to 0$ as $n \to \infty$ contradicting the continuity of A.

Remark. In the second part of the last proof, we did the following. To show that proposition P_1 (continuity) $\Rightarrow$ proposition P_2 (boundedness), we showed that the negation of $P_2 \Rightarrow$ the negation of P_1.

The greatest lower bound on all the numbers M for which (II.4-2) holds is called the *norm of the operator A* and is denoted $\|A\|$. The norm may be evaluated by any of the three equivalent expressions:

$$\|A\| = \sup_{\|x\| \leq 1} \|Ax\| = \sup_{\|x\| = 1} \|Ax\| = \sup_{x \neq 0} \frac{\|Ax\|}{\|x\|}. \tag{I.4-5}$$

The terminology of norm for $\|A\|$ as just defined is appropriate in the following context. With X and Y both normed linear spaces, let $L(X, Y)$ be the space of all bounded linear operators mapping X into Y. It is easily verified that $L(X, Y)$ is a linear space (with natural definitions for the algebraic operations) and, furthermore, $L(X, Y)$ is a normed linear space with the norm of (I.4-5) [i.e., (I.4-5) satisfies the conditions for a norm in Sec. I.3]. If Y is a Banach space (X need not be), then $L(X, Y)$ may be shown to be a Banach space.

Example I.4.1

Consider the mapping $y = Lx$ defined by the integral operator

$$y(t) = \int_a^b k(t, s)x(s)\, ds \qquad t \in [a, b], \tag{I.4-6}$$

where $k(t, s)$ is a continuous real-valued scalar function. The reader should verify that L is a linear transformation of $C(a, b)$ into itself. To show that it is a bounded linear operator, we have,

$$\|Lx\| = \max_{t \in [a,b]} \left| \int_a^b k(t, s)x(s)\, ds \right| \leq \|x\| \max_{t \in [a,b]} \int_a^b |k(t, s)|\, ds. \tag{I.4-7}$$

This implies that

$$\|L\| \leq \max_{t \in [a,b]} \int_a^b |k(t, s)|\, ds. \tag{I.4-8}$$

That the inequality of (I.4-8) may be replaced by equality may also be shown (Kantorovich and Akilov, p. 109) but not as easily as deriving (I.4-8). This situation is fairly typical: obtaining an M satisfying (I.4-2) is usually simple but verifying that the given M is the smallest such M requires more work.

We wish to distinguish between the mapping just discussed and $y = Lx$ defined by

$$y = \int_a^b k(s)x(s)\, ds \tag{I.4-9}$$

where $x \in C(a, b)$ and $k(s)$ is a real-valued integrable function. This last operator L maps $C(a, b)$ into R and hence is a linear functional. It is, in fact, a bounded linear functional (a bounded linear operator which is also a linear functional). A bound on its norm is also easily evaluated:

$$\|Lx\| = |y| \leq \int_a^b |k(s)x(s)|\, ds \leq \|x\| \int_a^b |k(s)|\, ds. \tag{I.4-10}$$

Let F map the set X onto the set Y and let it also be *one-one*, which means that, for any $y \in Y$, there is a unique $x \in X$ such that $y = Fx$. Then, for any $y \in Y$, the equation $y = Fx$ has a unique solution for x. We write this $x = F^{-1}(y)$ and call F^{-1} the *inverse* of F. It is clear that

$$F^{-1}F = I_X, \qquad FF^{-1} = I_Y,$$

where I_X and I_Y are the identity operators on X and Y, respectively.

The reader can easily verify that if F and G have inverses, then $(FG)^{-1} = G^{-1} F^{-1}$.

Exercise I.4.2

Show that if a linear operator has an inverse, its inverse must be linear.

The inverse of a linear operator is not necessarily a bounded linear operator (X and Y are now normed linear spaces). It may be shown that if A is a one-one onto bounded linear operator, then it has a bounded linear inverse (Porter, p. 433). The following two theorems are important in recognizing when a linear operator has a bounded inverse; they also give bounds on the norm of the inverse.

Theorem I.4.2

Assume that the equation

$$y = A(x) \qquad (A \text{ a linear operator}) \tag{I.4-12}$$

has a solution $x \in X$ for any $y \in Y$ and that

$$\|A(x)\| \geq m\|x\| \tag{I.4-13}$$

for all $x \in X$ and some $m > 0$. Then A^{-1} exists and is a bounded liner operator with

$$\|A^{-1}\| \leq \frac{1}{m}. \tag{I.4-14}$$

Proof. A is one-one since, if $x_1 \neq x_2$,

$$\|A(x_1) - A(x_2)\| = \|A(x_1 - x_2)\| \geq m\|x_1 - x_2\| > 0. \tag{I.4-15}$$

A is a mapping of X *onto* Y because of the solvability of (I.4-12). Hence, A^{-1} exists as a linear operator. Now,

$$\|y\| = \|A(x)\| \geq m\|x\| = m\|A^{-1}(y)\| \tag{I.4-16}$$

so that

$$\|A^{-1}(y)\| \leq \frac{1}{m} \|y\|. \tag{I.4-17}$$

Remarks. Note that the operator A is not required to be bounded. Also, do not confuse (I.4-13) with the definition of a norm of a linear operator (I.4-2); the inequalities are reversed. It is easily shown that the conditions are also necessary (as well as sufficient which was proved in Theorem I.4-2) for A to have a bounded inverse.

The next theorem, which we state without proof, is concerned with adding to the identity operator a bounded linear operator which is small in some sense. (It can be proved using Theorem II.1.1.)

Theorem I.4.3

Let X be a Banach space and $A \in L(X, X)$. If

$$\|A\| \leq \gamma < 1, \qquad (\text{I.4-18})$$

then the operator $I - A$ has a bounded linear inverse with

$$\|(I - A)^{-1}\| \leq \frac{1}{1 - \gamma}. \qquad (\text{I.4-19})$$

I.5 DERIVATIVES IN A BANACH SPACE

Most of the discussion of operators in the last section was concerned with linear operators. Nonlinear operators are, as to be expected, more difficult to work with. It is often useful to try to recapture some of the convenience of linear operators by using local approximation of nonlinear operators by linear operators. Derivatives in a Banach space do, in fact, just that. It is important that the reader keep in mind that derivatives are linear operators; this identification is not emphasized in ordinary calculus where the derivatives turn out to be numbers. However, there is a one-to-one correspondence between numbers and linear operators mapping the real line into itself (since such a linear operator L is representable as $Lx = kx$ with k a real number) so that the definition of derivatives to be given below represents a true generalization of the calculus definition.

Let F map an open subset Ω of a Banach space X into a Banach space Y. If $x_0 \in \Omega$ and if there is a bounded linear operator $F'(x_0) \in L(X, Y)$ such that for every $x \in X$

$$\lim_{\mu \to 0} \frac{F(x_0 + \mu x) - F(x_0)}{\mu} = F'(x_0)x, \qquad (\text{I.5-1})$$

then $F'(x_0)$ is called the *Gateaux derivative* of the operator F at the point x_0.

Note that, while F may only be defined on a subset of X, $F'(x_0)$ maps the whole space X into Y. In (I.5-1), $x_0 + \mu x \in \Omega$ for μ sufficiently small. If the convergence relationship of (I.5-1) is satisfied uniformly with respect to all $x \in X$ with $\|x\| = 1$, $F'(x_0)$ is called the *Fréchet derivative* of the operation F at x_0. By uniformity of the convergence we mean that the δ and ϵ in the definition of convergence can be selected independently of the x. Perhaps a more recognizable definition of Fréchet derivative is obtained with the equivalent representation if $\Delta x \neq 0$:

$$\lim_{\Delta x \to 0} \frac{\|F(x_0 + \Delta x) - F(x_0) - F'(x_0)\Delta x\|}{\|\Delta x\|} = 0. \tag{I.5-2}$$

When F has a Fréchet derivative at x_0, it is called *differentiable* at x_0. A Fréchet derivative is also obviously a Gateaux derivative but not vice versa. When we say derivative without specifying which one, we mean Gateaux derivative.

If F is itself a bounded linear operator, i.e., $F \in L(X, Y)$, F is differentiable at x_0 and

$$F'(x_0) = F \tag{I.5-3}$$

since

$$\frac{F(x_0 + \mu x) - Fx_0}{\mu} = \frac{Fx_0 + \mu Fx - Fx_0}{\mu} = Fx \tag{I.5-4}$$

for all $\mu \neq 0$.

Let $F = a_1 F_1 + a_2 F_2$ where a_1 and a_2 are scalars. If $F_1'(x_0)$ and $F_2'(x_0)$ exist, then so does $F'(x_0)$, and it equals $a_1 F_1'(x_0) + a_2 F_2'(x_0)$.

Now let P map the open set $\Omega \in X$ into the open set $\Delta \subset Y$ and let Q map Δ into Z (X, Y, and Z are Banach spaces). The composition mapping $R = QP$ maps Ω into Z and has a derivative at $x_0 \in \Omega$ (chain rule)

$$R'(x_0) = Q'(P(x_0))P'(x_0) = Q'(y_0)P'(x_0) \tag{I.5-5}$$

if Q is differentiable at $y_0 = P(x_0)$ and P has a derivative at x_0 (proof on p. 659 of Kantorovich and Akilov). For the special case of $Q = L$, a bounded linear operator, we have

$$R'(x_0) = LP'(x_0). \tag{I.5-6}$$

We now give some examples of derivatives. First we wish to show the familiar calculus definition is a special case.

Example I.5.1

Let $X = R$. If f maps an open set $\Omega \subset R$ into R, then the relationship (I.5-1) may be written (we can divide by a nonzero x since x is a number; we cannot do that in general)

$$\lim_{\mu \to 0} \frac{f(x_0 + \mu x) - f(x_0)}{\mu x} = f'(x_0). \tag{I.5-7}$$

Example I.5.2

Now let f map R^n into R^m where R^n and R^m are n-space and m-space, respectively (with any norms). That is, $y = f(x)$ with $y_i = f_i(x_1, \ldots, x_n)$, $i = 1, 2, \ldots, m$ (the subscripts refer to components of the vectors). It can be shown that if the partial derivatives $\partial f_i / \partial x_j$ exist in an open ball about $a \in R^n$ and are continuous at a, then the derivative operator exists at $x = a$ and is representable by a matrix (called the Jacobian matrix) whose ijth entry is $\partial f_i(a) / \partial x_j$. That is, $y = f'(a)x$ is given by

$$\begin{bmatrix} y_1 \\ \vdots \\ y_m \end{bmatrix} = \begin{bmatrix} \dfrac{\partial f_1(a)}{\partial x_1} & \cdots & \dfrac{\partial f_1(a)}{\partial x_n} \\ \vdots & & \vdots \\ \dfrac{\partial f_m(a)}{\partial x_1} & \cdots & \dfrac{\partial f_m(a)}{\partial x_n} \end{bmatrix} \begin{bmatrix} x_1 \\ \vdots \\ x_n \end{bmatrix}. \tag{I.5-8}$$

The derivative mapping is representable by a matrix as are all linear mappings of R^n into R^m.

Example I.5.3

Consider the integral operator $y = Fx$ defined by

$$y(t) = \int_a^b k(t, s, x(s)) \, ds. \tag{I.5-9}$$

Let $x_0 \in C(a, b)$ and $\Omega = \overline{S_r(x_0)}$ and also G is the set of points (s, t, u) in R^3 defined by $a \leq s, t \leq b$ and $|u - x_0(t)| \leq r$. If $k(t, s, u)$ is continuous and has a continuous derivative with respect to u at all points in G, then P maps Ω into $C(a, b)$ and is differentiable at every interior point $\bar{x} \in \Omega$, $y = F'(\bar{x})x$ being given by

$$y(t) = \int_a^b \frac{\partial k}{\partial u} (t, s, \bar{x}(s)) x(s) \, ds \tag{I.5-10}$$

(Kantorovich and Akilov, p. 680). In other words, the derivative operation is, loosely speaking, obtained by differentiating under the integral sign.

The following distinction is important. If F has a derivative at all $x_0 \in \Omega \subset X$, then $F'(x_0)$ is a bounded linear operator, i.e., $F'(x_0) \in L(X, Y)$. On the other hand we have the function F' or $F'(\cdot)$ mapping Ω into $L(X, Y)$. F' is not necessarily linear. To drive this point home, consider the function mapping R into itself, $f(x) = x^3$. The derivative operator $y = f'(x_0)x$ is given by $y = 3x_0^2 x$ which is linear in x but nonlinear in x_0.

I.6 SOME RECOLLECTIONS FROM CALCULUS; A NOTE ON MEASURABILITY

A real-valued function f of a real variable is recalled to be continuous at x_0 if $x \to x_0$ implies $f(x) \to f(x_0)$ (with x in the domain of f). A sequence of functions f_n defined on $[a, b]$ is said to *converge uniformly* to a function f defined on $[a, b]$ if for each $\epsilon > 0$ there is an integer $N(\epsilon)$ such that $n \geq N(\epsilon)$ implies $|f(x) - f_n(x)| < \epsilon$ for all $x \in [a, b]$. The essential point is that N be independent of x. As is known from advanced calculus, if all of the f_n are continuous, then so is f. Convergence in $C(a, b)$ is equivalent to uniform convergence. Hence a sequence of continuous functions which converges in the metric of $C(a, b)$ converges to a continuous function.

The function f is differentiable at x_0 if it is defined in an open set containing x_0 and

$$\lim_{x \to x_0} \frac{f(x) - f(x_0)}{x - x_0} \qquad (I.6-1)$$

exists; the limit is, of course, $f'(x_0)$. These definitions are special cases of the more general notions discussed previously in this chapter. Now let f be continuous on $[a, b]$ and differentiable at all $x \in (a, b)$. Then the *mean-value theorem* states that there is some $\theta \in (0, 1)$ such that

$$f(b) - f(a) = f'(a + \theta(b - a)) \cdot (b - a). \qquad (I.6-2)$$

If f is defined on $[a, b]$ and is continuous except at a finite number of points where it has limits from the left and the right (these are called jump discontinuities) f is *piecewise continuous*. If f is defined on $(-\infty, \infty)$, it is piecewise continuous on $(-\infty, \infty)$ if it is piecewise continuous on every finite interval.

Much of functional analysis uses a definition of integration (Lebesgue) more general than that of Riemann integration. Rather than even attempt to provide here some introduction to Lebesgue integration, we merely point out that, when the functions to be integrated are piecewise continuous, the two definitions of integration coincide. However, Lebesgue integration is applicable also to measurable functions which is a class of functions more general than the class of piecewise continuous functions. In fact, it is so general that practically every function one may think of is measurable. Those who are not presently familiar with measure and Lebesgue integration theory, when confronted with the few statements using measurability or Lebesgue terminology, might think in terms of piecewise continuous functions and Riemann integration with the realization that there is some loss of rigor (and they should put on their lists of things to do to learn some Lebesgue measure theory).

I.7 NOTES

Kantorovich and Akilov is the primary reference on functional analysis for this book. Some of their terminology has been modified to conform more with that of most American writers. In particular, they define a linear operator to be what we define to be a bounded linear operator. A briefer and more elementary reference is Kolmogorov and Fomin. The book by Porter complements the present book in that it gives a more detailed introduction to linear functional analysis with engineering motivations. Dieudonné may also be helpful in background reading. Further information on derivatives can be found in Nashed. A good introduction to Lebesgue integration theory is Natanson. We again mention that the terminology is not standardized so that the reader should always check out the definitions.

II

Use of the Contraction Mapping Theorem

This chapter will be concerned with proving existence of solutions to the equation

$$x = F(x). \tag{II-1}$$

An x satisfying (II-1) is called a *fixed point* of the operator F. The contraction mapping fixed point theorem is a widely used method of proving the existence of fixed points. The theorem is the basis for much of the analysis of this book. It is simple to prove and often simple to use. The purpose of this chapter is to develop some facility in its use so that it will be a readily available tool in later chapters. As an example of the theorem's usefulness, it can provide information on how an exact solution differs from an approximate solution.

II.1 THE CONTRACTION MAPPING THEOREM (CMT)

Let X be a complete metric space (with metric d) containing the closed set Ω and let F map Ω into itself. F is said to satisfy a *Lipschitz condition* with

Lipschitz constant γ if

$$d(F(x), F(x')) \le \gamma \, d(x, x') \qquad \text{(all } x, x' \in \Omega). \qquad \text{(II.1-1)}$$

If $\gamma < 1$, then F is a *contraction* on Ω and γ is a *contraction constant*.

Theorem II.1.1 (CMT)

If F is a contraction mapping on Ω, then there is a unique fixed point x^* of F in Ω and

$$x^* = \lim_{n \to \infty} x_n = \lim_{n \to \infty} F^n(x_0) \qquad \text{(II.1-2)}$$

where

$$x_{n+1} = F(x_n) \qquad n = 0, 1, \ldots \qquad \text{(II.1-3)}$$

and x_0 is any element of Ω. Furthermore,

$$d(x^*, x_n) \le \frac{\gamma^n \, d(x_1, x_0)}{1 - \gamma}. \qquad \text{(II.1-4)}$$

Proof. Let $x_0 \in \Omega$ and consider the sequence defined by (II.1-3), Then with the integer $p \ge 1$ and using the triangle inequality,

$$\begin{aligned}
d(x_{n+p}, x_n) &= d(F^{n+p}(x_0), F^n(x_0)) \le \gamma^n \, d(F^p(x_0), x_0) = \gamma^n \, d(x_p, x_0) \\
&\le \gamma^n [d(x_p, x_{p-1}) + d(x_{p-1}, x_{p-2}) \\
&\qquad + \cdots + d(x_2, x_1) + d(x_1, x_0)] \\
&\le \gamma^n [\gamma^{p-1} + \gamma^{p-2} + \cdots + \gamma + 1] \, d(x_1, x_0) \\
&\le \frac{\gamma^n \, d(x_1, x_0)}{1 - \gamma}. \qquad \text{(II.1-5)}
\end{aligned}$$

Since $\lim_{n \to \infty} \gamma^n = 0$, it is seen that $\{x_n\}$ is a Cauchy sequence and since X is a complete space, $x_n \to x^* \in X$. Since every $x_n \in \Omega$ and Ω is closed, $x^* \in \Omega$. The operation F is continuous by virtue of being a contraction mapping so that

$$x^* = \lim_{n \to \infty} x_n = \lim_{n \to \infty} F(x_{n-1}) = F(\lim_{n \to \infty} x_{n-1}) = F(x^*). \qquad \text{(II.1-6)}$$

To show that x^* is unique in Ω, assume that there is another fixed point $x^{**} \in \Omega$. Then

$$d(x^*, x^{**}) = d(F(x^*), F(x^{**})) \le \gamma \, d(x^*, x^{**}). \qquad \text{(II.1-7)}$$

Since $\gamma < 1$, we must have $d(x^*, x^{**}) = 0$ or $x^* = x^{**}$. The inequality of (II.1-4) is obtained by letting $p \to \infty$ in (II.1-5).

Remarks.

(i) The condition (II.1-1) cannot be replaced by $d(F(x), F(x')) < d(x, x')$ unless further conditions are imposed (a counterexample is given in Kantorovich and Akilov, p. 629).

(ii) From (II.1-4) we obtain

$$d(x^*, x_0) \le \frac{d(x_1, x_0)}{1 - \gamma}. \tag{II.1-8}$$

If x_0 is an approximate solution, then (II.1-8) bounds the error in approximation.

(iii) For a graphical interpretation of the CMT, see Kolmogorov and Fomin, p. 44.

Example II.1.1

Consider the scalar differential equation

$$\frac{dx(t)}{dt} = f(x(t), t), \qquad x(a) = c. \tag{II.1-9†}$$

Is there a solution to this differential equation in an interval containing a? An affirmative answer can be given provided we put appropriate conditions on f. Let $f(u, t)$ be defined for $u \in U = \{u : u \in R, |u - c| \le \delta\}$, $t \in [a, b]$ and let it be continuous in t for each fixed $u \in U$. Furthermore, assume that

$$|f(u, t)| \le M, \qquad u \in U, \qquad t \in [a, b] \tag{II.1-10}$$

and

$$|f(u_1, t) - f(u_2, t)| \le k|u_1 - u_2|, \qquad u_1, u_2 \in U, \qquad t \in [a, b] \tag{II.1-11}$$

where k is independent of t.

Now note that a solution to (II.1-9) is equivalent to a solution of the integral equation

$$x(t) = c + \int_a^t f(x(s), s) \, ds. \tag{II.1-12}$$

---

† At ends of intervals, derivatives are one-sided.

We show that (II.1-12) has a solution in $C(a, b)$, the space of continuous functions on $[a, b]$, if

$$b - a < \min \left\{ \frac{\delta}{M}, \frac{1}{k} \right\}. \tag{II.1-13}$$

The norm on $C(a, b)$ is recalled to be

$$\|x\| = \max_{t \in [a,b]} |x(t)|. \tag{II.1-14}$$

Let

$$\Omega = \{x : x \in C(a, b), \quad \max_{t \in [a,b]} |x(t) - c| \leq \delta\}. \tag{II.1-15}$$

Ω is a closed subset of the Banach space $C(a, b)$. Consider the operator $z = Fx$ defined by

$$z(t) = c + \int_a^t f(x(s), s) \, ds. \tag{II.1-16}$$

F maps Ω into itself because z is clearly continuous from (II.1-16), and if $x \in \Omega$ then $z = Fx$ satisfies

$$|z(t) - c| \leq \int_a^t |f(x(s), s)| \, ds \leq M(b - a) < \delta, \quad t \in [a, b]. \tag{II.1-17}$$

Now we show that F is a contraction on Ω. If $x_1, x_2 \in \Omega$,

$$\begin{aligned}
\|F(x_1) - F(x_2)\| &= \max_{t \in [a,b]} \left| \int_a^t [f(x_1(s), s) - f(x_2(s), s)] \, ds \right| \\
&\leq \max_{t \in [a,b]} \int_a^t k |x_1(s) - x_2(s)| \, ds \\
&\leq k(b - a) \|x_1 - x_2\|. \tag{II.1-18}
\end{aligned}$$

Thus, from (II.1-13), $k(b - a)$ is a contraction constant and the conditions of the CMT are satisfied.

Exercise II.1.1

Generalize the preceding example to a system of differential equations:

$$\frac{dx_i(t)}{dt} = f_i(x_1(t), \ldots, x_n(t), t), \quad i = 1, \ldots, n. \tag{II.1-19}$$

A defect of the result of the preceding example is that existence is guaranteed only on the interval of length $(b - a)$ which depends on the bound in (II.1-10) and Lipschitz constant in (II.1-11) and which may turn out to be quite small. A stronger type of result may be obtained by using the corollary to the following theorem.

Theorem II.1.2

Let F map a nonempty set S into itself. Let K be an invertible function mapping S onto S. Then the function F has a fixed point if and only if the composite function $K^{-1}FK$ has a fixed point.

Proof. If y is a fixed point of $K^{-1}FK$, then $y = K^{-1}FKy$ and $Ky = FKy$ so that Ky is a fixed point of F. Similarly, if x is a fixed point of F, $y = K^{-1}x$ is a fixed point of $K^{-1}FK$.

Corollary

Let F map the complete metric space X into itself. Let there exist an invertible mapping K of X onto itself. Then if $K^{-1}FK$ is a contraction, F has a unique fixed point.

Proof. Since $K^{-1}FK$ has a unique fixed point by the CMT, then so does F by Theorem II.1.2.

Example II.1.2

Let us apply this corollary to showing existence of a unique solution to the scalar differential equation

$$\frac{dx(t)}{dt} = f(x(t), t), \qquad x(a) = c, \qquad t \in [a, b], \tag{II.1-20}$$

where $f(u, t)$ is continuous in t for any $u \in R$ and it satisfies the Lipschitz condition

$$|f(u_1, t) - f(u_2, t)| \leq k|u_1 - u_2|, \qquad u_1, u_2 \in R, \qquad t \in [a, b], \tag{II.1-21}$$

and b is *any* finite number bigger than a. The mapping $z = Fx$ defined by (II.1-16) maps $C(a, b)$ into itself. The mapping $z = Kx$ defined by

$$z(t) = e^{\lambda k(t-a)}x(t), \qquad \lambda > 1, \qquad t \in [a, b], \tag{II.1-22}$$

also maps $C(a, b)$ into itself and has an inverse defined by the function $e^{-\lambda k(t-a)}$. To show that $K^{-1}FK$ is a contraction, let $y_1, y_2 \in C(a, b)$ and then

$$
\begin{aligned}
\|K^{-1}FKy_1 - K^{-1}FKy_2\| &= \max_{t \in [a,b]} \left| e^{-\lambda k(t-a)} \int_a^t \left[f(e^{\lambda k(s-a)} y_1(s), s) \right. \right. \\
&\qquad\qquad \left. \left. - f(e^{\lambda k(s-a)} y_2(s), s) \right] ds \right| \\
&\leq \max_{t \in [a,b]} e^{-\lambda k(t-a)} \int_a^t k e^{\lambda k(s-a)} |y_1(s) - y_2(s)| \, ds \\
&\leq \max_{t \in [a,b]} \lambda^{-1}(1 - e^{-\lambda k(t-a)}) \|y_1 - y_2\| \\
&\leq \lambda^{-1} \|y_1 - y_2\|. \quad\quad\quad\quad\quad\quad\quad\quad\quad \text{(II.1-23)}
\end{aligned}
$$

Since $\lambda > 1$, $K^{-1}FK$ is a contraction.

Remark. The preceding example shows that the successive approximations $y_{n+1} = K^{-1}FKy_n$ converge to $y^* = K^{-1}FKy^*$ and $x^* = Ky^*$ is a fixed point of F. This also implies the convergence of the successive approximations $x_{n+1} = Fx_n$ to x^*. To see this, let

$$
x_{n+1} = Ky_{n+1} = FKy_n = Fx_n. \quad\quad\quad\quad \text{(II.1-24)}
$$

We see from the continuity of K (it is a bounded linear operator on the finite interval $[a, b]$; see Theorem 1.4.1) the convergence of y_n implies convergence of x_n. Furthermore, using the relation [from (II.1-4)]

$$
\|y^* - y_n\| \leq \frac{(\lambda)^{-n} \|K^{-1}FKy_0 - y_0\|}{1 - (\lambda)^{-1}} \quad\quad\quad\quad \text{(II.1-25)}
$$

we find that

$$
\|K^{-1}(x^* - x_n)\| \leq \frac{(\lambda)^{-n} \|K^{-1}(Fx_0 - x_0)\|}{1 - (\lambda)^{-1}} \quad\quad\quad\quad \text{(II.1-26)}
$$

which implies that for $t \in [a, b]$

$$
|x^*(t) - x_n(t)| \leq \frac{(\lambda)^{-n} e^{\lambda k(t-a)}}{1 - (\lambda)^{-1}} \max_{t \in [a,b]} |e^{-\lambda k(t-a)}[Fx_0 - x_0](t)|. \quad \text{(II.1-27)}
$$

Exercise II.1.2

Show that the same type of result as obtained in the last example may be obtained by changing the norm on the Banach space to

$$||x|| = \max_{t \in [a,b]} |e^{-\lambda k(t-a)} x(t)| \tag{II.1-28}$$

and not modifying the map F.

The last example shows that while F itself may not be a contraction, some modification of it may be and this may be sufficient to prove the existence of a fixed point. Another possibility is that in the event F is not a contraction, F^k may be a contraction for some positive integer k and this gives existence of a fixed point.

Theorem II.1.3

F maps the complete metric space X into itself. If, for some positive integer k, F^k is a contraction, then F has a unique fixed point.

Proof. Denote by x^* the unique fixed point of F^k. Since

$$F(x^*) = F(F^k(x^*)) = F^{k+1}(x^*) = F^k(F(x^*)), \tag{II.1-29}$$

it follows that $F(x^*)$ is a fixed point of F^k and so $F(x^*) = x^*$. Thus F has a fixed point and a fixed point of F is necessarily a fixed point of F^k and so is unique.

Remark. Note that F being a contraction implies that F is continuous but if F^k is a contraction only for some $k > 1$, then F may not be continuous. For example let F be defined by

$$F(x) = \begin{cases} 0 & x \in [0, 1] \\ 1 & x \in (1, 2]. \end{cases} \tag{II.1-30}$$

Then $F^2(x) = 0$ for all $x \in [0, 2]$ and F^2 is a contraction on $[0, 2]$, but F is not continuous.

Exercise II.1.3

Apply the last theorem to the Volterra integral equation

$$x(t) = \lambda \int_a^t K(t, s)x(s)\, ds + \phi(t), \qquad t \in [a, b], \tag{II.1-31}$$

where $K(t, s)$ and $\phi(t)$ are continuous functions.

We have seen that substantial profit may be derived by appropriately modifying the operators whose fixed point is sought. Next, we consider a very simple example which contains the germ of the idea in the main result of the next chapter.

Example II.1.3

Consider finding a fixed point of $-f(x)$ with f a real-valued differentiable function on $(-\infty, \infty)$. If $|f'(x)| \leq \gamma < 1$ for all x, then using the mean-value theorem (Sec. I.6.1),

$$|-f(b) - (-f(a))| \leq \sup_{\theta \in (0,1)} |f'(a + \theta(b - a))| \cdot |b - a| \leq \gamma |b - a|,$$

$$(II.1\text{-}32)$$

we see that $(-f)$ is a contraction and thus has a unique fixed point. But suppose that $\alpha \leq f'(x) \leq \beta$ with $\beta > 1$. Let us manipulate the equation as follows:

$$x = -[f(x) - cx] - cx \qquad (c \neq -1),$$
$$x = -(1 + c)^{-1}[f(x) - cx] \qquad (II.1\text{-}33)$$
$$\equiv m(x).$$

and try to find conditions for m to be a contraction [a fixed point of m is clearly a fixed point of $(-f)$]. We have

$$|m'(x)| \leq |1 + c|^{-1} \max \{|\beta - c|, |c - \alpha|\}. \qquad (II.1\text{-}34)$$

The components of the right-hand side of the last inequality are plotted in Fig. II.1. It is seen that for $\alpha > -1$ the minimum of the right-hand side is taken on at $c = \frac{1}{2}(\alpha + \beta)$ in which case

$$|1 + c|^{-1} \max \{|\beta - c|, |c - \alpha|\} = \frac{\beta - \alpha}{2 + \beta + \alpha} < 1. \qquad (II.1\text{-}35)$$

In the next chapter, we show that this example can be surprisingly generalized to the equation $x = -LNx + r$ in an arbitrary Banach space with N a nonlinear operator and with L a quite general linear operator (in the above example, L is merely the identity operator and $r = 0$).

We are often interested in fixed points which depend on parameters (e.g., parameters in a differential equation). That is, if $x = F(x, y)$ then x depends on y, $x = x(y)$. The next theorem is concerned with continuous dependence on y. The theorem uses the notion of product set defined in Sec. I.1 (product spaces are discussed in more detail in Sec. VII.1).

Theorem II.1.4

X and Y are metric spaces, X is complete, and Ω is a closed subset of X. F maps $\Omega \times Y$ into itself. For every $y \in Y$, for all $x, x' \in \Omega$,

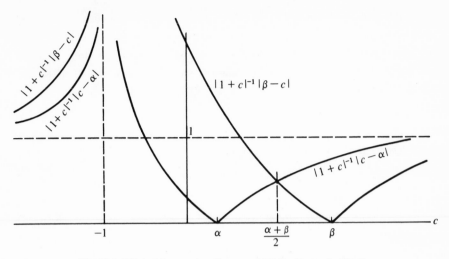

Fig. II.1 Minimizing contraction constant for Example II.1.3

$$d(F(x, y), F(x', y)) \leq \gamma \, d(x, x') \qquad \text{(II.1-36)}$$

with $\gamma < 1$ and γ independent of y. Also, for every $x \in \Omega$, F is continuous with respect to y at $y_0 \in Y$. Then the fixed point $x = F(x, y)$ is continuous with respect to y at y_0.

Proof. From Theorem II.1.1, for any $y \in Y$, $F(x, y)$ has a unique fixed point $x^*(y)$ in Ω satisfying

$$d(x^*(y), x_0) \leq \frac{d(F(x_0, y), x_0)}{1 - \gamma}. \qquad \text{(II.1-37)}$$

Letting $x_0 = F(x^*(y_0), y_0) = x^*(y_0)$, we have

$$d(x^*(y), x^*(y_0)) \leq \frac{d[F(x^*(y_0), y), F(x^*(y_0), y_0)]}{1 - \gamma} \qquad \text{(II.1-38)}$$

and the conclusion follows from the continuity of F with respect to y at y_0.

Remark. If, for every $x \in \Omega$ and for all $y, y' \in Y$,

$$d(F(x, y), F(x, y')) \leq \gamma_1 \, d(y, y') \qquad \text{(II.1-39)}$$

then, from (II.1-38),

$$d(x^*(y), x^*(y_0)) \leq \frac{\gamma_1 \, d(y, y_0)}{1 - \gamma}. \qquad \text{(II.1-40)}$$

II.2 USE OF DERIVATIVES

In Example II.1.3, the mean value theorem for real functions of a real variable was used. The mean-value theorem (see Sec. I.6) states that under certain conditions there is a $\theta \in (0, 1)$ such that

$$f(b) - f(a) = f'(a + \theta(b - a))(b - a). \tag{II.2-1}$$

In Example II.1.3, we used the following implication of (II.2-1):

$$|f(b) - f(a)| \leq \sup_{\theta \in (0,1)} |f'(a + \theta(b - a))| \cdot |b - a|. \tag{II.2-2}$$

Thus f satisfies a Lipschitz condition which is clearly relevant to contraction mappings.

The generalization of derivatives in Sec. I.5 leads to a relation analogous to (II.2-2) but not (II.2-1). That (II.2-1) is not generally extendable may be seen by considering the function of a complex variable, $f(x) = e^x$. Let $a = 0, b = 2\pi i$. Then $f(2\pi i) - f(0) = 0$ but there is no $\theta \in (0, 1)$ such that $f'(\theta 2\pi i) = 0$. The proof of the following theorem is given in the appendix to this chapter.

Theorem II.2.1

Let $x_0, x \in \Omega$, an open subset of a Banach space X, let $\mu x_0 + (1 - \mu)x \in \Omega$ for $\mu \in [0, 1]$, and let F have a derivative at all these points. Then, with $\Delta x = x - x_0$,

$$\|F(x) - F(x_0)\| \leq \sup_{\theta \in (0,1)} \|F'(x_0 + \theta\Delta x)\| \cdot \|\Delta x\|. \tag{II.2-3}$$

Now we have the following theorem.

Theorem II.2.2

Let F map an open subset Ω_0 of a Banach space B into B and have a derivative at every point of a convex closed set $\Omega \subset \Omega_0$. Then if F maps Ω into itself and

$$\sup_{x \in \Omega} \|F'(x)\| = \gamma < 1, \tag{II.2-4}$$

there exists a unique fixed point of F in Ω.

Proof. Let $x_1, x_2 \in \Omega$. From (II.2-3)

$$\|F(x_2) - F(x_1)\| \leq \sup_{\theta \in (0,1)} \|F'(x_1 + \theta(x_2 - x_1))\| \cdot \|x_2 - x_1\|$$

$$\leq \gamma \|x_2 - x_1\| \tag{II.2-5}$$

and thus the conditions of the CMT (Theorem II.1.1) are satisfied.

Remark. The convexity requirement is imposed on Ω so that $x_1 + \theta(x_2 - x_1)$ $\in \Omega$ if $x_1, x_2 \in \Omega$ [one of the assumptions used in deriving Eq. (II.2-3)].

Note that this last theorem is less general than the CMT in Sec. II.1 in that derivatives are assumed to exist. However, if the derivatives do exist, much benefit may be derived since we may be able to work with a specific function with the Lipschitz constant information associated with it. This will become more evident in Sec. II.3.

II.3 FINDING THE APPROPRIATE SET MAPPED INTO ITSELF

The CMT is simplest to use when Ω is the entire metric space X. This also, of course, imposes the restriction that $d(F(x_1), F(x_2)) \leq \gamma\, d(x_1, x_2)$ for $\gamma < 1$ and all $x_1, x_2 \in X$, which may not be met in many cases. Then one can look for a closed subset Ω mapped into itself such that F is a contraction on that set. When $\Omega \neq X$, i.e., it is a proper subset, we shall refer to F as a local contraction if F is a contraction on Ω. When $\Omega = X$, we speak of a global contraction. This terminology is somewhat artificial since a closed subset of a complete metric space may be considered to be a metric space by itself and is then complete (Sec. I.2). We shall nevertheless find the terminology useful to distinguish between the cases where we consider a mapping on a whole space (the original one) and where we go searching for an appropriate subset Ω mapped into itself.

Let us consider the problem of searching for Ω. Say that x_0 is an "approximate" fixed point of F, i.e., $d(F(x_0), x_0)$ is "not too large." To find a fixed point near x_0, we desire that $d(F(x_1), F(x_2)) \leq \gamma\, d(x_1, x_2)$ for x_1, x_2 near x_0. To be more concrete, let

$$\gamma = \sup_{\substack{x_1, x_2 \in \bar{S}_r(x_0) \\ x_1 \neq x_2}} \frac{d(F(x_1), F(x_2))}{d(x_1, x_2)}$$

Now γ is actually a function of r; in fact, it is a nondecreasing function of r so that we wish to choose r as small as possible. On the other hand, r must be large enough so that the ball is mapped into itself. In particular, it must be at least as large as $d(F(x_0), x_0)$. We can be even more definite as to the size of r. From (II.1-8) it is seen that if F is a contraction on Ω, then by an iteration starting at x_0, we can conceivably go as far away as $d(F(x_0), x_0)/(1 - \gamma)$. We thus have to find a contraction constant γ on a ball whose radius depends on γ. The following theorem is motivated by the preceding discussion.

Theorem II.3.1

Let F map Ω_0, a subset of complete metric space X, into itself and let $x_0 \in \Omega_0$. Assume that $\gamma < 1$ and $d(F(x_1), F(x_2)) \leq \gamma \, d(x_1, x_2)$ for all x_1, x_2 in the ball $\overline{S_r(x_0)} \subset \Omega_0$ and that $d(F(x_0), x_0) \leq r(1 - \gamma)$. Then there is a unique fixed point of F in $\overline{S_r(x_0)}$.

Proof. We only have to show that the ball is mapped into itself for the conditions of the CMT to be satisfied. To this end, let $d(x, x_0) \leq r$. Then

$$
\begin{aligned}
d(F(x), x_0) &\leq d(F(x), F(x_0)) + d(F(x_0), x_0) \\
&\leq \gamma \, d(x, x_0) + r(1 - \gamma) \\
&\leq \gamma r + r(1 - \gamma) \\
&= r.
\end{aligned}
\tag{II.3-1}
$$

Remarks.

(i) There may be other fixed points outside $\overline{S_r(x_0)}$.

(ii) The fixed point x^* satisfies $d(x^*, x_0) \leq d(F(x_0), x_0)/(1 - \gamma)$; see Theorem II.1.1. A sharper estimate on the fixed point may be obtained from (II.1-4) with $n > 0$.

(iii) Sometimes a modification of Theorem II.3.1 is presented which requires that $d(F(x'), F(x)) \leq \gamma \, d(x', x)$ in a ball defined by $d(x, x_1) \leq (\gamma/1 - \gamma) \, d(x_1, x_0)$ with $x_1 = F(x_0)$ (see, e.g., Saaty, p. 56 or Collatz, pp. 213, 214). This can potentially give a sharper result than Theorem II.3.1 if the Lipschitz constants are estimated on this ball (centered at x_1) by means better than using the Lipschitz constants on a ball with center at x_0 and the triangle inequality.

The above theorem actually formalizes the discussion preceding it but finding the appropriate r and γ still may be difficult in some cases because it may be complicated to relate Lipschitz constants to balls of radius r. The following theorem is less general but is often convenient in finding the appropriate r and γ.

Theorem II.3.2

Let F map an open subset Ω_0 of a Banach space B into B and let F have a derivative at every point of Ω_0. Then if $x_0 \in \Omega_0$ and if there is a nondecreasing function g such that

$$\|F'(x)\| \leq g(\|x - x_0\|), \qquad x \in \Omega_0, \qquad \text{(II.3-2)}$$

and

$$g\left(\frac{k}{1 - \gamma}\right) \leq \gamma < 1, \qquad \text{(II.3-3)}$$

where

$$k \geq \|F(x_0) - x_0\|, \qquad \text{(II.3-4)}$$

and if

$$\Omega = \left\{x : x \in B, \|x - x_0\| \leq \frac{k}{1 - \gamma}\right\} \subset \Omega_0, \qquad \text{(II.3-5)}$$

then there is a unique fixed point of F in Ω.

Proof. We show that the conditions of Theorem II.2.2 are satisfied. If $x \in \Omega$,

$$\|F'(x)\| \leq g(\|x - x_0\|) \leq g\left(\frac{k}{1 - \gamma}\right) \leq \gamma < 1.$$

To show that Ω is mapped into itself, let $x \in \Omega$ and using (II.2-3),

$$\begin{aligned}
\|F(x) - x_0\| &\leq \|F(x) - F(x_0)\| + \|F(x_0) - x_0\| \\
&\leq \gamma\|x - x_0\| + k \\
&\leq \frac{\gamma k}{1 - \gamma} + k \\
&= \frac{k}{1 - \gamma}.
\end{aligned}$$

Remarks.

(i) That Theorem II.3.2 is a special case of Theorem II.3.1 may be seen by letting $r = k/(1 - \gamma)$. Then condition (II.3-3) may be restated as $g(r) \leq 1 - k/r$.

(ii) Often, $\Omega_0 = B$ so that Ω is automatically contained in Ω_0.

(iii) Again, a sharper estimate on the fixed point may be obtained by using (II.1-4) with $n > 0$.

(iv) It is not actually necessary for F to have a derivative if g is interpreted as a function relating Lipschitz constants to radii r of balls centered at x_0. That is,

$$g(r) = \sup_{\substack{x_1, x_2 \in S_r(x_0) \\ x_1 \ne x_2}} \frac{\|F(x_1) - F(x_2)\|}{\|x_1 - x_2\|}. \tag{II.3-6}$$

But it is the case when F has a derivative that $g(r)$ is most easily evaluated so that Theorem II.3.2 becomes very convenient.

(v) It is often a simple matter to find an appropriate g function. For example, if $\|F'(x)\| \le h(\|x\|)$ where h is a nondecreasing function, then we can use $g(\|x - x_0\|) = h(\|x_0\| + \|x - x_0\|)$ via the triangle inequality. If h is a nonincreasing function, then $h(|\|x_0\| - \|x - x_0\||)$ is a g function candidate [using (I.3-3)].

Example II.3.1

Consider the function of a real variable

$$f(x) = x^3 \tag{II.3-7}$$

which obviously has fixed points at $x = 0$ and $x = 1$. Let us apply Theorem II.3.2. We have

$$|f'(x)| = 3x^2 \le 3(|x_0| + |x - x_0|)^2 \equiv g(|x - x_0|) \tag{II.3-8}$$

$$k = |x_0^3 - x_0|. \tag{II.3-9}$$

Condition (II.3-3) is satisfied if

$$3\left(|x_0| + \frac{|x_0^3 - x_0|}{1 - \gamma}\right)^2 \le \gamma < 1. \tag{II.3-10}$$

For $x_0 = 0.1$, (II.3-10) is satisfied with $\gamma = 0.14$ and the fixed point x^* satisfies

$$|x^* - x_0| \le \frac{k}{1 - \gamma} = 0.115. \tag{II.3-11}$$

Of course, in this trivial example it is known that the fixed point $x^* = 0$ actually statisfies $|x^* - x_0| = 0.1$. In more realistic examples (e.g., in Chapter V), the exact information is unavailable.

Exercise II.3.1

Apply Theorem II.3.2 to the function of a real variable, $f(x) = \frac{1}{2}\pi \sin x$. Let $x_0 = 85°$. (The following bound can be obtained: $|x^* - x_0| \le 0.120 = 7°$.)

A mapping F which is not a global contraction can have two distinct fixed points. Then the possibility arises of using the CMT in neighborhoods of each fixed point. The following theorem indicates when this cannot be done. It will be found useful when we discuss jump resonance (which is concerned with distinct fixed points) in Sec. V.5.

Theorem II.3.3

Let F map an open convex subset Ω of a Banach space into itself and let $x_1 = F(x_1)$ and $x_2 = F(x_2)$ with $x_1, x_2 \in \Omega$, $x_1 \neq x_2$, $\|x_2\| \geq \|x_1\|$. Suppose F has a derivative at every point of Ω and that $\|F'(x)\| \leq h(\|x\|)$ for all $x \in \Omega$ where h is a nondecreasing function. Then $h(\|x_2\|) \geq 1$.

Proof. From Theorem II.2.1 and using the triangle inequality

$$\|x_1 - x_2\| = \|F(x_1) - F(x_2)\| \leq \sup_{\theta \in (0,1)} \|F'(x_1 + \theta(x_2 - x_1))\| \cdot \|x_1 - x_2\|$$

$$\leq \sup_{\theta \in (0,1)} h(\|x_1 + \theta(x_2 - x_1)\|) \cdot \|x_1 - x_2\|$$

$$\leq \sup_{\theta \in (0,1)} h[(1 - \theta)\|x_1\| + \theta\|x_2\|] \cdot \|x_1 - x_2\|$$

$$\leq h(\|x_2\|)\|x_1 - x_2\|$$

which implies $h(\|x_2\|) \geq 1$.

This result may be used in two ways. If $\|F'(x)\| = h(\|x\|)$, then the mapping itself evaluated in a neighborhood of x_2 is not a contraction. If h is a function bounding the norm of F' which is being used in trying to apply the CMT, then the attempt will be unsuccessful with x_2. One might then try to sharpen the bound on $\|F'(x)\|$ in a neighborhood of x_2.

Exercise II.3.2

How does Theorem II.3.3 relate to Example II.3.1 and Exercise II.3.1?

II.4 INTERPRETATION AND RELATION TO APPROXIMATIONS

Let us note that the condition $g(k/1 - \gamma) \leq \gamma$ in Theorem II.3.2 may be satisfied by more than one γ. The smallest such γ gives the sharpest information on the fixed point while a larger γ gives a larger ball in which the fixed point is unique. The following may be helpful in visualizing the application of

Theorem II.3.2. Assume that the condition of Theorem II.3.2 is satisfied with equality, that is,

$$g\left(\frac{k}{1-\gamma}\right) = \gamma. \qquad \text{(II.4-1)}$$

The radius of the ball Ω is $k/(1-\gamma)$ [see Remark (i) to Theorem II.3.2]. Letting

$$r = \frac{k}{1-\gamma} \qquad \text{(II.4-2)}$$

the condition is seen to be

$$g(r) = 1 - \frac{k}{r} \qquad \text{(II.4-3)}$$

which Fig. II.2 shows pictorially.

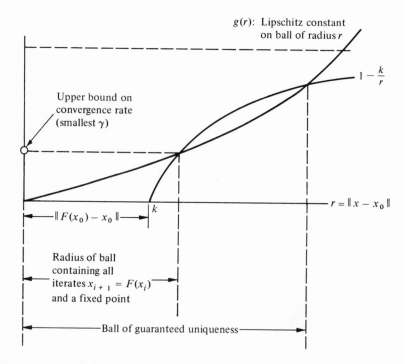

Figure II.2 Contraction mapping theorem interpretation

Many approximate methods (e.g., describing functions) may be viewed in the context of minimizing k. The error estimate $\|x^* - x_0\| \leq k/(1-\gamma) = r$

clearly depends on k (which in turn depends on x_0). There is thus some reasonableness in choosing x_0 to make k as small as possible: It tends to minimize a bound on the error between the approximate and exact solutions (x_0 and x^*, respectively). It does not necessarily minimize the actual error but the error bound is often the only information available.

We wish to clarify what we mean by choosing x_0 to make k small since, if we choose x_0 to be x^*, the actual fixed point (if one exists), then k is clearly zero. However, we usually do not know x^* (otherwise this chapter would be unnecessary) so we might choose x_0 subject to some constraint. In particular, we could choose x_0 as the solution of an approximate equation

$$x_0 = \tilde{F}(x_0), \tag{II.4-4}$$

where $\tilde{F}$ is related in some manner to F, but it is easier to find fixed points of $\tilde{F}$. We shall find that some commonly used approximation techniques indeed can be explained in the above context.

Let us note that γ also depends on x_0. In this connection, assume that $\tilde{F}$ has two fixed points, x_{01} and x_{02} with $\|x_{02}\| > \|x_{01}\|$. Also assume that

$$\|F'(x)\| \leq h(\|x\|), \tag{II.4-5}$$

where h is a nondecreasing function. As mentioned in Remark (v) to Theorem II.3.2, a candidate for the function g is

$$g(\|x - x_0\|) = h(\|x_0\| + \|x - x_0\|). \tag{II.4-6}$$

Let g_1 and g_2 denote (II.4-6) associated with x_{01} and x_{02}, respectively. Then, in view of Theorem II.3.3, a possible situation is depicted in Fig. II.3, where k_1

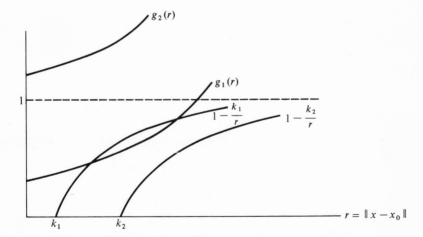

Figure II.3 Nonunique fixed points

$= \|F(x_{01}) - x_{01}\|$ and $k_2 = \|F(x_{02}) - x_{02}\|$. For Fig. II.3, it is assumed that there are fixed points x_1^* and x_2^* of F near x_{01} and x_{02}, respectively, with $\|x_2^*\| \geq \|x_1^*\|$.

II.5 DISCUSSION

Though this book primarily uses the contraction mapping fixed point theorem, the reader should be made aware that there are other fixed point theorems which may be applicable when the mapping is not a contraction. The reason for the primary use of the CMT here is that, while there are many cases in which the mapping is not a contraction, we have seen that certain simple manipulations can repose the problem so that the CMT becomes applicable. Thus the CMT actually becomes applicable (directly or indirectly) to a rather wide class of problems. The CMT is also relatively unsophisticated mathematically compared to some other fixed point theorems and the reader, after not a too lengthy mathematical preparation, can apply the CMT. Furthermore, the CMT, as contrasted with some other fixed point theorems, provides an iterative method for constructing solutions with explicit error bounds [see (II.1-3) and (II.1-4)].

The preceding is not meant to imply that we wish to discourage the reader from investigating other fixed point theorems. On the contrary, we try to clearly indicate defects and limitations of the CMT analyses. This book will surely have served its purpose if some readers are stimulated to overcome these defects, perhaps by using other fixed point theorems (which are discussed in many books on functional analysis, e.g., Edwards, Kantorovich and Akilov).

We also note that not only are there fixed point theorems not of the contractive type but there are also other generalizations of the CMT (some are discussed in Collatz).

Since we are confining this book to the CMT let us discuss how to get the most mileage out of it. The CMT states that, if F maps a closed subset Ω of a complete metric space into itself, and if

$$\sup_{\substack{x, x' \in \Omega \\ x \neq x'}} \frac{d(F(x), F(x'))}{d(x, x')} = \gamma \tag{II.5-1}$$

with $\gamma < 1$, then there is a unique fixed point of F in Ω. If $\gamma \geq 1$, several things might be tried to make the CMT applicable. First of all, one might change the metric (see Exercise II.1-2). We are, however, not aware of any theory or general guidelines to use in this connection. An example of a contraction mapping analysis carried out with three different metrics is given in Kolmogorov and Fomin, Vol. I, pp. 45,46. For this example, there are cases where any one of the metrics results in a contraction while the other two do not.

In this chapter, we have discussed two other methods of reducing γ to below unity. The first is to modify the mapping in such a manner so that the modified mapping is a contraction and the fixed point of the modified map is also a fixed point of the original map. The corollary to Theorem II.1.2, Theorem II.1.3, and Example II.1.3 use this approach. The second method of reducing γ in (II. 5-1) is to make Ω as small as possible (which is the idea behind Theorem II.3.2).

The "best" result (if we measure "best" by smallest γ)† would presumably be obtained by using some combination of the above approaches. However, the best result obtainable by any one approach is an area not yet completely resolved. For example, is is not known what is the best function K to use in Theorem II.1.2. Note that in Example II.1.2, the contraction constant can be made arbitrarily small. On the other hand, the best choice of c in Example II.1.3 was found. The next chapter discusses a class of operator equations (which include as special cases the equations of many feedback systems). The problem of minimizing the contraction constant, in a certain context, is completely solved and the solution turns out to be most simple. In Sec. VII. 4, we shall briefly discuss Newton's method in the context of minimizing a contraction constant.

II.6 NOTES

Fixed point theorems are discussed in many places, e.g., in Kantorovich and Akilov, Edwards, Collatz, Dieudonné, Saaty, and a chapter by Wouk in Anselone. Theorem II.1.2 and its corollary are taken from Chu and Diaz. The proof of Theorem II.1.3 is taken from Bryant. Theorem II.1.4 is from pp. 630 and 631 of Kantorovich and Akilov. Theorem II.3.2 and its interpretation are based on Holtzman [1] and [2].

---

† See the Remark to Example II.1.2, which relates estimates on the original equation to the contraction constant λ^{-1}.

APPENDIX

Proof of Theorem II.2.1

We use the following result stated without proof.

Theorem II.A

If X is a normed linear space and x_0 is a nonzero element of X, a linear functional exists such that

$$\|f\| = 1 \qquad \text{and} \qquad f(x_0) = \|x_0\|. \tag{II.A-1}$$

Remark. If X is a Hilbert space (to be discussed in Sec. IV.1), we can set

$$f(x) = \langle x, x_0/\|x_0\| \rangle.$$

Proof of Theorem II.2.1. Define for $\mu \in [0, 1]$

$$\phi(\mu) = g(F(x_0 + \mu\Delta x)), \tag{II.A-2}$$

where g is the linear functional on X satisfying (using Theorem II.A)

$$\|g\| = 1, \qquad g(F(x_0 + \Delta x) - F(x_0)) = \|F(x_0 + \Delta x) - F(x_0)\|, \qquad \Delta x \neq 0. \tag{II.A-3}$$

If X is a complex space, ϕ is a complex-valued function of the real variable μ. To show that $\phi'(\mu)$ exists for $\mu \in (0, 1)$, we write

$$\frac{\phi(\mu + \Delta\mu) - \phi(\mu)}{\Delta\mu} = g\left(\frac{F(x_0 + \mu\Delta x + \Delta\mu\,\Delta x) - F(x_0 + \mu\Delta x)}{\Delta\mu}\right).$$
$$(\text{II.A-4})$$

When $\Delta\mu \to 0$, the argument of g has limit $F'(x_0 + \mu\Delta x)\,\Delta x$ and since g is a bounded linear functional, hence continuous (Theorem I.4.1),

$$\phi'(\mu) = g(F'(x_0 + \mu\Delta x)\Delta x).\qquad(\text{II.A-5})$$

The function ϕ satisfies

$$|\phi(1) - \phi(0)| \leq \sup_{\theta \in (0,1)} |\phi'(\theta)| \cdot |1 - 0|\qquad(\text{II.A-6})$$

or

$$|g(F(x_0 + \Delta x) - F(x_0))| \leq \sup_{\theta \in (0,1)} |g(F'(x_0 + \theta\Delta x)\Delta x)|.\qquad(\text{II.A-7})$$

Using (II.A-3) yields

$$\|F(x_0 + \Delta x) - F(x_0)\| \leq \sup_{\theta \in (0,1)} \|F'(x_0 + \theta\Delta x)\| \cdot \|\Delta x\|.\qquad(\text{II.A-8})$$

Remark. The above proof is based on Kantorovich and Akilov, pp. 659, 660, except that they use a real linear functional g for a real normed linear space in which case (II.A-6) follows from the mean-value theorem for real-valued functions,

$$\phi(1) - \phi(0) = \phi'(\theta),\qquad(\text{II.A-9})$$

[see (I.6-2)]. With ϕ complex-valued, we do not have (II.A-9), but (II.A-6) can be verified.

III

Contraction Mapping Analysis

of

Nonlinear Feedback Systems

In this chapter, we consider an equation which describes many nonlinear feedback systems. A simple technique for minimizing the contraction constant (in a certain context) is presented. The main result is a far-reaching generalization of Example II.1.3. In an important case when the feedback system consists of a zero-memory nonlinearity (not necessarily time-invariant; i.e., $y = Nx$ is defined by $y(t) = n(x(t), t)$) in cascade with a linear element representable by a frequency response transfer function, a most interesting geometrical interpretation may be given to the conditions for a contraction (which implies existence of a unique response to a given input). The interpretation is much like Nyquist's criterion.

III.1 THE SYSTEM TO BE STUDIED

Consider the system described by the equation

$$x = -LNx + r, \qquad (III.1-1)$$

where L is a linear operator mapping a real Banach space B into itself and N

maps B into itself and is such that there are two real constants α and β ($\beta \geq \alpha$ and $\beta > 0$) with the property that for any real constant c and all $x_1, x_2 \in B$,

$$\|Nx_1 - cx_1 - (Nx_2 - cx_2)\| \leq \eta(c)\|x_1 - x_2\|, \qquad \text{(III.1-2)}$$

where

$$\eta(c) = \max\{|\beta - c|, |c - \alpha|\}. \qquad \text{(III.1-3)}$$

The block diagram representation of (III.1-1) shown in Fig. III.1 shows that (III.1-1) represents a large class of feedback systems. By choosing different Banach spaces (e.g., a space of periodic functions, a space of random functions, etc.), the results of this chapter become very widely applicable.

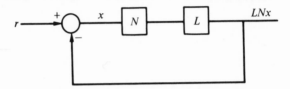

Figure III.1 Block diagram of feedback system

Equation (III.1-1) is sometimes called a Hammerstein equation. It also describes many systems not necessarily described from the feedback point of view.

At the end of this chapter, we shall give an example of (III.1-1) for a specific Banach space. Until then, we will just let B be an arbitrary Banach space. However, since it may not be obvious at first glance what nonlinearities can be described by (III.1-2) and (III.1-3), consider for the moment a nonlinear function mapping $L_2(-\infty, \infty)$ into itself where $L_2(-\infty, \infty)$ is the space of real-valued (measurable) functions on $(-\infty, \infty)$ satisfying

$$\int_{-\infty}^{\infty} |x(t)|^2 \, dt < \infty. \qquad \text{(III.1-4)}$$

It may be shown that $L_2(-\infty, \infty)$ is a Banach space with the norm

$$\|x\|^2 = \int_{-\infty}^{\infty} |x(t)|^2 \, dt. \qquad \text{(III.1-5)}$$

Assume the nonlinear function $y = Nx$ is defined by $y(t) = n(x(t), t)$ for $t \in (-\infty, \infty)$ and

(i) $n(x(t), t)$ is measurable when $x(t)$ is   (III.1-6)
(ii) $n(0, t) = 0 \qquad t \in (-\infty, \infty)$ (III.1-7)
(iii) there are two real numbers α and β such that

$$\alpha \leq \frac{n(u_1, t) - n(u_2, t)}{u_1 - u_2} \leq \beta, \qquad u_1 \neq u_2, \qquad t \in (-\infty, \infty). \quad \text{(III.1-8)}$$

Condition (iii) imposes upper and lower bounds on the slope of the nonlinear function n. At any fixed $t \in (-\infty, \infty)$ the function $n(u, t) - cu$ (mapping R into itself) has a slope with magnitude not exceeding $\max \{|\beta - c|, |c - \alpha|\}$. Then, with $x_1, x_2 \in L_2(-\infty, \infty)$

$$\|Nx_1 - cx_1 - (Nx_2 - cx_2)\|^2 = \int_{-\infty}^{\infty} |n(x_1(t), t) - cx_1(t) - (n(x_2(t), t)$$

$$- cx_2(t))|^2 \, dt$$

$$\leq (\max \{|\beta - c|, |c - \alpha|\})^2 \int_{-\infty}^{\infty} |x_1(t)$$

$$- x_2(t)|^2 \, dt$$

$$= (\max \{|\beta - c|, |c - \alpha|\})^2 \|x_1 - x_2\|^2$$

$$(\text{III.1-9})$$

so that (III.1-2) is satisfied.

We can try to apply the CMT to the mapping on the right-hand side of (III.1-1) but we shall find it fruitful to first modify the mapping to one whose fixed point (if it exists) is also a fixed point of the original mapping. Let the operator $\tilde{N}$ be defined by

$$Nx = cx + \tilde{N}x. \qquad (\text{III.1-10})$$

Then (III.1-1) becomes

$$x = -L(cx + \tilde{N}x) + r. \qquad (\text{III.1-11})$$

Now, if $(I + cL)^{-1}$ exists,

$$x = -(I + cL)^{-1} L\tilde{N}x + (I + cL)^{-1} r \qquad (\text{III.1-12})$$

(see Fig. III.2). It is clear that solutions of (III.1-12) coincide with solutions of (III.1-1). We shall use the CMT on (III.1-12) and find the c which minimizes the contraction constant.

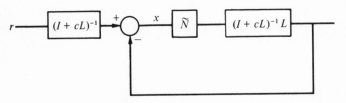

Figure III.2 Transformed system

The following theorem is a straightforward application of the CMT.

Theorem III.1.1

Let c be a real number such that $(I + cL)^{-1}$ exists and suppose that $\gamma(c)$ $= \|(I + cL)^{-1}\|\eta(c) < 1$. Then for any $r \in B$, there exists a unique $x \in B$ satisfying (III.1-1). Furthermore, $x = \lim\limits_{n \to \infty} x_n$ where

$$x_{n+1} = -(I + cL)^{-1}L\tilde{N}x_n + (I + cL)^{-1}r. \qquad \text{(III.1-13)}$$

Note that this theorem may be applicable even though the CMT might be inapplicable if applied to (III.1-1). That is, if $c = 0$, the mapping defined by the right-hand side of (III.1-13) might not be a contraction. With $c = 0$, the CMT would require, in essence, that the "open loop incremental gain" be less than unity, a condition so restrictive that it is seldom met in practice. The question that immediately comes to mind is what is the "best" c to use. This question has a surprisingly simple answer as we show in the next section.

III.2 MINIMIZING THE CONTRACTION CONSTANT

Define D as the set of real c such that $(I + cL)^{-1}$ exists, i.e.,

$$D = \{c : c \in R, (I + cL)^{-1} \text{ exists}\}. \qquad \text{(III.2-1)}$$

Now if there is a c such that

$$\gamma(c) < 1, \qquad \text{(III.2-2)}$$

where $\gamma(c)$ is defined in the Theorem III.1.1, we will show that

$$\gamma(c_0) = \inf_{c \in D} \gamma(c), \qquad \text{(III.2-3)}$$

where

$$c_0 = \tfrac{1}{2}(\alpha + \beta). \qquad \text{(III.2-4)}$$

Note that this result is independent of the choice of Banach space or linear operators. For the simple Example II.1.3, the best c was found to be the average of the maximum and minimum slopes of the function $(-f)$. This is a special case of the general result of this chapter.

We state this result as the following theorem.

Theorem III.2.1

Suppose there exists a $c \in D$ such that $\gamma(c) < 1$. Then $c_0 = \tfrac{1}{2}(\alpha + \beta) \in D$ and $\gamma(c_0) \leq \gamma(c)$.

Proof. We first show that $(I + c_0L)^{-1}$ exists. The inverse obviously exists if $L = 0$ (the null operator). If $L \neq 0$ and $c \in D$ then $\|(I + cL)^{-1}L\| \neq 0$. From

$$I + c_0L = I + cL + (c_0 - c)L$$
$$= (I + cL)[I + (c_0 - c)(I + cL)^{-1}L], \qquad (III.2-5)$$

it is clear that $(I + c_0L)^{-1}$ exists if

$$|c_0 - c| \cdot \|(I + cL)^{-1}L\| < 1 \qquad (III.2-6)$$

(see Theorem I.4.3). But this condition is met because $|c_0 - c| \leq \eta(c)$ [see Fig. III.3(a)] and $\gamma(c) < 1$. Using the fact that $\eta(c_0) = \eta(c) - |c - c_0|$ [see Fig. III.3(b)],

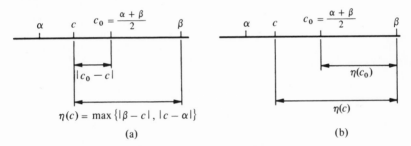

$$\eta(c) = \max\{|\beta - c|, |c - \alpha|\}$$

(a)

$$\eta(c_0)$$

$$\eta(c)$$

(b)

Figure III.3 Illustration of steps in proof of Theorem III.2.1

$$\|(I + c_0L)^{-1}L\|\eta(c_0) = \|(I + c_0L)^{-1}L\|\eta(c) - |c - c_0| \cdot \|(I + c_0L)^{-1}L\|$$
$$\leq \|(I + cL)^{-1}L\|\eta(c) + \|(I + c_0L)^{-1}L$$
$$- (I + cL)^{-1}L\|\eta(c)$$
$$- |c - c_0| \cdot \|(I + c_0L)^{-1}L\|. \qquad (III.2-7)$$

(The last step used the triangle inequality.) Since

$$(I + c_0L)^{-1} - (I + cL)^{-1} = (I + c_0L)^{-1}[(I + cL) - (I + c_0L)](I + cL)^{-1}$$
$$= (c - c_0)(I + c_0L)^{-1}L(I + cL)^{-1} \qquad (III.2-8)$$

we have

$$\|(I + c_0L)^{-1}L - (I + cL)^{-1}L\|\eta(c) = |c - c_0| \cdot \|(I + c_0L)^{-1}L(I + cL)^{-1}L\|\eta(c)$$
$$\leq |c - c_0| \cdot \|(I + c_0L)^{-1}L\| \cdot \|(I + cL)^{-1}L\|\eta(c)$$
$$= |c - c_0| \cdot \|(I + c_0L)^{-1}L\|\gamma(c). \qquad (III.2-9)$$

Thus, from (III.2-7) and (III.2-9)

$$\gamma(c_0) = \|(I + c_0 L)^{-1} L\| \eta(c_0) \le \gamma(c) - (1 - \gamma(c))|c - c_0| \cdot \|(I + c_0 L)^{-1} L\|$$

$$(\text{III.2-10})$$

and since $\gamma(c) < 1$ we have $\gamma(c_0) \le \gamma(c)$.

Exercise II.2.1

Verify the statement at the beginning of the proof that $\|(I + cL)^{-1} L\| \ne 0$.

III.3 FOURIER TRANSFORMS

In the next section, we consider feedback systems of the form of (III.1-1) where the linear operator is described by its frequency response transfer function, i.e., the Fourier transform of an impulse response. For purposes of reference, we first give here some properties of Fourier transforms most of which we state without proof.

For $1 \le p < \infty$, and $a < b$, $L_p(a, b)$ is defined here as the space of all real-valued (measurable) functions f for which $\int_a^b |f(t)|^p \, dt < \infty$. $L_p(a, b)$ is a Banach space with norm

$$\|f\|_p = \left(\int_a^b |f(t)|^p \, dt \right)^{1/p}. \qquad (\text{III.3-1})$$

The numbers a and b may be finite or infinite.† Two L_p spaces of immediate interest here are $L_1(-\infty, \infty)$ and $L_2(-\infty, \infty)$. If $f, g \in L_2(-\infty, \infty)$ then the product of f and g (not composition) $fg \in L_1(-\infty, \infty)$ and it statisfies the *Schwarz inequality*,

$$\|fg\|_1 \le \|f\|_2 \cdot \|g\|_2, \qquad (\text{III.3-2})$$

i.e.,

$$\left(\int_{-\infty}^{\infty} |f(t)g(t)| \, dt \right)^2 \le \int_{-\infty}^{\infty} |f(t)|^2 \, dt \int_{-\infty}^{\infty} |g(t)|^2 \, dt. \qquad (\text{III.3-3})$$

First we define the *Fourier transform of an $L_1(-\infty, \infty)$ function.* If $f \in L_1(-\infty, \infty)$, its Fourier transform $F(i\omega)$ is defined by

$$F(i\omega) = \int_{-\infty}^{\infty} f(t) e^{-i\omega t} \, dt \qquad (\text{III.3-4})$$

† The functions are defined on $[a, b]$ if a and b are finite, on $[a, \infty)$ if $b = \infty$, and on $(-\infty, \infty)$ if $b = -a = \infty$.

and exists for all real ω. Some properties of $F(i\omega)$ are immediate from some elementary properties of (Lebesgue) integrals:

(i) $F(i\omega)$ is bounded:

$$\sup_{\omega \in (-\infty, \infty)} |F(i\omega)| \leq \int_{-\infty}^{\infty} |f(t)e^{-i\omega t}| \, dt = \int_{-\infty}^{\infty} |f(t)| \, dt = \|f\|_1. \quad \text{(III.3-5)}$$

(ii) $F(i\omega)$ is uniformly continuous:

$$|F(i(\omega + h)) - F(i\omega)| \leq \int_{-\infty}^{\infty} |e^{-iht} - 1| \cdot |f(t)| \, dt. \quad \text{(III.3-6)}$$

The integral on the right-hand side of (III.3-6) approaches zero as $h \to 0$.

Since $f \in L_1(-\infty, \infty)$ is real-valued, the real and imaginary parts of $F(i\omega)$ are easily seen to be even and odd, respectively. That is, $\text{Re}[F(-i\omega)] = \text{Re}[F(i\omega)]$ and $\text{Im}[F(-i\omega)] = -\text{Im}[F(i\omega)]$. Hence, $\overline{F(i\omega)} = F(-i\omega)$.

The *Riemann-Lebesgue lemma* states that the Fourier transform of an $L_1(-\infty, \infty)$ function must approach zero as $\omega \to \pm\infty$, i.e.,

$$\lim_{\omega \to \pm\infty} F(i\omega) = 0. \quad \text{(III.3-7)}$$

A natural question to ask is: given $F(i\omega)$, the Fourier transform of an $L_1(-\infty, \infty)$ function f, what is f? Though this question has been answered, its answer is more complicated than the same question for $L_2(-\infty, \infty)$ functons. Since we shall need the inversion theorem only for $L_2(-\infty, \infty)$ functions we postpone discussion of inversion until $L_2(-\infty, \infty)$ functions are treated. It may suffice to point out that a candidate for an inversion formula for $f \in L_1 (-\infty, \infty)$,

$$f(t) = \frac{1}{2\pi} \int_{-\infty}^{\infty} F(i\omega)e^{i\omega t} \, d\omega, \quad \text{(III.3-8)}$$

is not adequate in general since $f \in L_1(-\infty, \infty)$ does not imply that $F(i\omega)$ is integrable so the integral on the right-hand side of (III.3-8) may not even exist. Equation (III.3-8) is valid sometimes, e.g., when both $f(t)$ is continuous at t and $F(i\omega)$ is absolutely integrable over $(-\infty, \infty)$.

If $f, g \in L_1(-\infty, \infty)$ then the integral

$$h(t) = \int_{-\infty}^{\infty} f(t - u)g(u) \, du \quad \text{(III.3-9)}$$

exists for all t (more precisely, almost everywhere†) and $h \in L_1(-\infty, \infty)$. The

---

† A statement is true almost everywhere (a.e.) if it is true except on a set of zero measure which is a negligible set in the sense that the integral of a measurable function over a zero measure set is zero. We shall not always distinguish between functions that differ only on a zero measure set because they are "essentially" the same (the norm of the difference is zero if the norm integrates the difference).

function h is called the *convolution* of f and g and the following notation is used:

$$h = f * g. \tag{III.3-10}$$

Some properties of convolution are listed below. The functions $f, g, k \in L_1$ $(-\infty, \infty)$ and $H(i\omega)$, $F(i\omega)$, $G(i\omega)$ are the respective Fourier transforms of h, f, g.

$$f * g = g * f \tag{III.3-11}$$

$$(f * g) * k = f * (g * k) \tag{III.3-12}$$

$$\|f * g\|_1 \leq \|f\|_1 \cdot \|g\|_1 \tag{III.3-13}$$

$$h = f * g \Rightarrow H(i\omega) = F(i\omega)G(i\omega) \tag{III.3-14}$$

Property (III.3-14) is most important and states that convolving f and g corresponds to multiplying their Fourier transforms. This is the basis of the widely used transfer function concept..

Now suppose that $f \in L_2(-\infty, \infty)$. We should note that $f \in L_2(-\infty, \infty)$ does not imply nor is implied by $f \in L_1(-\infty, \infty)$. It may be instructive for the reader to show by example the truth of this assertion. If $|b - a| < \infty$, then $f \in L_2(a, b)$ implies that $f \in L_1(a, b)$. The *Fourier transform of an $L_2(-\infty, \infty)$ function* is defined by

$$F(i\omega) = \underset{N \to \infty}{\text{l.i.m.}} \, F_N(i\omega) \tag{III.3-15}$$

where $F_N(i\omega)$ is the $L_1(-\infty, \infty)$ Fourier transform of f_N which is defined by

$$f_N(t) = f(t) \qquad (|t| \leq N)$$

$$f_N(t) = 0 \qquad (|t| > N) \tag{III.3-16}$$

and where l.i.m. (limit in the mean) is defined by

$$\lim_{N \to \infty} \int_{-\infty}^{\infty} |F(i\omega) - F_N(i\omega)|^2 \, dt = 0 \tag{III.3-17}$$

(this limit may be shown to exist). Note that since $f \in L_2(-\infty, \infty)$ the restriction of f to $[-N, N]$ (i.e., f considered only on $[-N, N]$) is a member of $L_2(-N, N)$ and hence, by the remark above, also a member of $L_1(-N, N)$. This further implies that $f_N \in L_1(-\infty, \infty)$ and thus has a Fourier transform as defined for $L_1(-\infty, \infty)$ functions.

In contrast to the remark on inversion of the Fourier transform of an $L_1(-\infty, \infty)$ function, the Fourier transform of an $L_2(-\infty, \infty)$ function is square integrable on $(-\infty, \infty)$. We do not say here that it is in $L_2(-\infty, \infty)$ because the transform may be complex-valued and $L_2(-\infty, \infty)$ was defined here to consist only of real-valued functions. In some references, $L_2(-\infty, \infty)$ is defined to include complex-valued functions.

We list below some important properties of Fourier transforms of $L_2(-\infty, \infty)$ functions. Both $f, g \in L_2(-\infty, \infty)$ with $F(i\omega)$, $G(i\omega)$ their respective transforms and overbar here denotes complex conjugate.

(i) $\displaystyle\int_{-\infty}^{\infty} f(t)g(t)\, dt = \frac{1}{2\pi} \int_{-\infty}^{\infty} F(i\omega)\overline{G(i\omega)}\, d\omega.$ (III.3-18)

(ii) Parseval's relation [special case of (i)]:

$$\|f\|_2 = \frac{1}{\sqrt{2\pi}}\left(\int_{-\infty}^{\infty} |F(i\omega)|^2\, d\omega\right)^{1/2}.$$ (III.3-19)

(iii) Inversion formula:

$$f(t) = \underset{N\to\infty}{\text{l.i.m.}} \frac{1}{2\pi} \int_{-N}^{N} F(i\omega)e^{i\omega t}\, d\omega.$$ (III.3-20)

Remark. If $F(i\omega)$ is square integrable over $(-\infty, \infty)$, and if $\overline{F(i\omega)} = F(i\omega)$ $\overset{-i\omega}{}$ (for realness of f), then $F(i\omega)$ is the Fourier transform of the unique $L_2(-\infty, \infty)$ function f given by (III.3-20).

For $f, g \in L_2(-\infty, \infty)$, we have, analogously to the case for $L_1(-\infty, \infty)$ functions, that the Fourier transform of the convolution of f and g is the product of the respective transforms. One can also define the convolution of $f \in L_2(-\infty, \infty)$ with $g \in L_1(-\infty, \infty)$. In that case, relation (III.3-14) remains valid with $h \in L_2(-\infty, \infty)$ and the Fourier transform of h is the product of $F(i\omega)$ and $G(i\omega)$. Note that since $G(i\omega)$ is bounded [see (III.3-5)] and $F(i\omega)$ is square integrable, then it follows that $H(i\omega)$ is square integrable [as it should be, being the Fourier transform of an $L_2(-\infty, \infty)$ function; see (III.3-19)]. Even more generally, if $f \in L_p(-\infty, \infty)$ and $g \in L_1(-\infty, \infty)$, $p \geq 1$, we find that $f * g$ exists and

$$\|f * g\|_p \leq \|g\|_1 \cdot \|f\|_p.$$ (III.3-21)

In Chapter VIII, we shall use Fourier transforms of $L_1(0, \infty)$ and $L_2(0, \infty)$ functions. We can use the definitions of (III.3-4) and (III.3-15) if the functions are assumed zero for $t < 0$, i.e., by changing the limits of integration to non-negative times.

III.4 A CIRCLE CONDITION

We now apply the results of Sec. III.1 and III.2 to an important case where the Banach space is $L_2(-\infty, \infty)$. Our objective is to obtain a frequency domain interpretation of the minimum contraction constant condition ($\gamma(c_0) < 1$) which resembles Nyquist's criterion.

The operation N is defined by the real-valued function of two real variables, $n(u, t)$, which satisfies the following properties

(i) $n(x(t), t)$ is measurable when $x(t)$ is measurable.   (III.4-1)

(ii) $n(0, t) = 0$ for $t \in (-\infty, \infty)$.   (III.4-2)

(iii) $\alpha \leq \dfrac{n(u_1, t) - n(u_2, t)}{u_1 - u_2} \leq \beta$ for all $u_1 \neq u_2, t \in (-\infty, \infty)$.   (III.4-3)

This is the nonlinear function mentioned in Sec. III.1 which satisfies (III.1-2).

The linear operator $y = Lx$ is defined by the convolution operation

$$y(t) = \int_{-\infty}^{\infty} h(t - u)x(u)\,du \qquad\qquad \text{(III.4-4)}$$

with $h \in L_1(-\infty, \infty)$. In many cases of engineering interest, $h(t) = 0$ for $t < 0$, but this restriction is not necessary for the present analysis. That L maps $L_2(-\infty, \infty)$ into itself has already been pointed out near the end of Sec. III.3.[†] The Fourier transform of h, $H(i\omega)$, is assumed to satisfy

$$\tfrac{1}{2}(\alpha + \beta)H(i\omega) \neq -1, \qquad \omega \in (-\infty, \infty). \qquad \text{(III.4-5)}$$

We will show that the condition for minimum contraction constant (minimum in the context of Sec. III.2) is

$$\sup_{\omega \in (-\infty, \infty)} \left| \frac{H(i\omega)}{1 + \tfrac{1}{2}(\alpha + \beta)H(i\omega)} \right| \tfrac{1}{2}(\beta - \alpha) < 1. \qquad \text{(III.4-6)}$$

Note that $(\beta - \alpha)$ might be regarded as a measure of departure from linearity. When $\beta = \alpha$, the contraction constant is zero.

Condition (III.4-6) has a most interesting geometric interpretation:

(i) When $\alpha > 0$, the locus of $H(i\omega)$ for $\omega \in (-\infty, \infty)$ lies outside the circle C_1 of radius $\tfrac{1}{2}(\alpha^{-1} - \beta^{-1})$ centered in the complex plane at $[-\tfrac{1}{2}(\alpha^{-1} + \beta^{-1}), 0]$.

(ii) When $\alpha = 0$, $\text{Re}[H(i\omega)] > -\beta^{-1}$ for all real ω.

(iii) When $\alpha < 0$, the locus of $H(i\omega)$ for $\omega \in (-\infty, \infty)$ is contained within the circle C_2 of radius $\tfrac{1}{2}(\beta^{-1} - \alpha^{-1})$ centered in the complex at $[-\tfrac{1}{2}(\alpha^{-1} + \beta^{-1}), 0]$.

Note that with the above interpretations, the loci of $H(i\omega)$ are bounded away from the forbidden regions, i.e., they come no closer than some positive number δ.

---

† It should also be noted that a necessary and sufficient condition for L to map bounded functions x into bounded functions y is that $h \in L_1(-\infty, \infty)$ (see, e.g., Zadeh and Desoer, p. 401).

Exercise III.4.1

Verify conditions (i)–(iii).

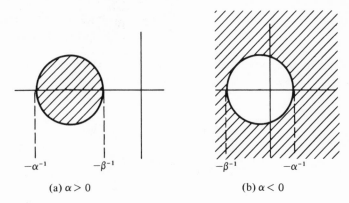

(a) $\alpha > 0$ (b) $\alpha < 0$

Figure III.4 Forbidden regions in the complex plane

Figure III.4 shows the circles C_1 and C_2. The cross hatched areas show the forbidden regions of the complex plane that the locus of $H(i\omega)$ for $\omega \in (-\infty, \infty)$ must not enter if the contraction condition (III.4-6) is to be satisfied. It should be noted that for nonlinear feedback control systems this circle condition is rather restrictive. In particular, if the nonlinearity is a saturation or a deadzone function,† then $\alpha = 0$ and we must have $\text{Re}[H(i\omega)] > -\beta^{-1}$, a condition often violated. The question then arises as to how far from a necessary condition the circle condition is (it was derived as a sufficient condition for the existence of unique responses). We shall discuss this question in a closely related context in our discussion of periodic inputs (Sec. IV.5). Uniqueness in a space of periodic functions implies absence of jump resonance.

To derive (III.4-6), we must show how to work with norms and inverses in $L_2(-\infty, \infty)$. It follows from (III.3-21) that

$$\|y\| \leq \int_{-\infty}^{\infty} |h(t)| \, dt \, \|x\| \tag{III.4-7}$$

[the norm is for $L_2(-\infty, \infty)$] which implies that the norm of the linear operator satisfies

$$\|L\| \leq \int_{-\infty}^{\infty} |h(t)| \, dt. \tag{III.4-8}$$

† The continuous function $n(\sigma)$ is a saturation [deadzone] function if $n(\sigma) = \beta\sigma$ for $|\sigma| \leq a$ and $n(\sigma) = \beta a \operatorname{sgn} \sigma$ for $|\sigma| > a$ [$n(\sigma) = 0$ for $|\sigma| \leq a$, $n(\sigma) = \beta(\sigma - a)$ for $\sigma > a$ and $n(\sigma) = \beta(\sigma + a)$ for $\sigma < -a$].

However, the right-hand side of (III.4-8) is only a bound on $\|L\|$; it is not, in general, equal to $\|L\|$. Instead, we have

$$\|L\| = \sup_{\omega \in (-\infty, \infty)} |H(i\omega)|$$

$$= \sup_{\omega \in [0, \infty)} |H(i\omega)|, \qquad (III.4\text{-}9)$$

the last equality due to the realness of h which implies $\overline{H(i\omega)} = H(-i\omega)$. To verify (III.4-9), let $f \in L_2(-\infty, \infty)$ and $g = Lf$; then with $F(i\omega)$, the Fourier transform of f,

$$\|g\|^2 = \frac{1}{2\pi} \int_{-\infty}^{\infty} |H(i\omega)|^2 |F(i\omega)|^2 \, d\omega$$

$$\leq \sup_{\omega \in (-\infty, \infty)} |H(i\omega)|^2 \|f\|^2, \qquad (III.4\text{-}10)$$

showing that

$$\|L\| \leq \sup_{\omega \in (-\infty, \infty)} |H(i\omega)|. \qquad (III.4\text{-}11)$$

Now note that, since $\lim_{\omega \to \pm\infty} H(i\omega) = 0$ (Riemann-Lebesgue lemma), and from the continuity of $|H(i\omega)|$, that there must be an ω_0 such that[†]

$$|H(i\omega_0)| = \sup_{\omega \in (-\infty, \infty)} |H(i\omega)|. \qquad (III.4\text{-}12)$$

Consider the sequence of functions $f_n \in L_2(-\infty, \infty)$ with Fourier transforms

$$F_n(i\omega) = \sqrt{\pi n}, \qquad \omega \in \left[\omega_0 - \frac{1}{2n}, \omega_0 + \frac{1}{2n}\right],$$

$$\omega \in \left[-\omega_0 - \frac{1}{2n}, -\omega_0 + \frac{1}{2n}\right]$$

$$= 0, \text{ otherwise.} \qquad (III.4\text{-}13)$$

The $F_n(i\omega)$ are transforms of the functions $f_n(t) = \sqrt{\pi n}\, 2 \cos \omega_0 t \sin (1/2n)t/(\pi t)$; $\|f_n\| = 1$ (see Papoulis, pp. 20, 122). With $g_n = Lf_n$

$$\left| |H(i\omega_0)|^2 - \|g_n\|^2 \right| = \left| \int_{\omega_0 - 1/2n}^{\omega_0 + 1/2n} n(|H(i\omega_0)|^2 - |H(i\omega)|^2) \, d\omega \right|$$

$$\leq \max_{\omega \in [\omega_0 - 1/2n,\ \omega_0 + 1/2n]} \left| |H(i\omega_0)|^2 - |H(i\omega)|^2. \right| \quad (III.4\text{-}14)$$

[†] This follows from the result from advanced calculus that a continuous function takes on its maximum in a nonempty closed, bounded interval and by considering the interval $[-w, w]$ with $|H(i\omega)| < M$ for $|\omega| > w$ and $\sup_{\omega} |H(i\omega)| = M > 0$.

Condition (III.4-14) shows that by choosing n large enough, $\|g_n\|$ can be made arbitrarily close to $|H(i\omega_0)|$. Since $\|g_n\| \leq \|L\|$, this implies that the inequality of (III.4-11) may be replaced by equality.

We now show that (III.4-5) is equivalent to

$$\inf_{\omega \in (-\infty, \infty)} |1 + \tfrac{1}{2}(\alpha + \beta)H(i\omega)| > 0 \qquad \text{(III.4-15)}$$

because of the continuity of $H(i\omega)$ and the Riemann–Lebesgue lemma. That is, for (III.4-15) not to hold, there must be a sequence $\{\omega_n\}$ and an ω_0 such that

$$\lim_{\omega_n \to \omega_0} \tfrac{1}{2}(\alpha + \beta)H(i\omega_n) = -1. \qquad \text{(III.4-16)}$$

Since $\lim_{\omega \to \pm \infty} H(i\omega) = 0$, ω_0 must be finite. But the continuity of $H(i\omega)$ implies $\lim_{\omega \to \omega_0} \tfrac{1}{2}(\alpha + \beta)H(i\omega) = \tfrac{1}{2}(\alpha + \beta)H(i\omega_0) = -1$ which contradicts (III.4-5). We use (III.4-15) in the following.

To use the results of Sec. III.1 and III.2, we must first show that $(I + \tfrac{1}{2}(\alpha + \beta)L)$ has an inverse. Let $f \in L_2(-\infty, \infty)$ and

$$g = (I + \tfrac{1}{2}(\alpha + \beta)L)f. \qquad \text{(III.4-17)}$$

With $G(i\omega)$, $F(i\omega)$, $H(i\omega)$ the Fourier transforms of g, f, h, we have

$$\|g\|^2 = \frac{1}{2\pi} \int_{-\infty}^{\infty} \left| \left(1 + \frac{\alpha + \beta}{2} H(i\omega)\right) F(i\omega)\right|^2 d\omega$$

$$\geq \inf_{\omega \in (-\infty, \infty)} \left|1 + \frac{\alpha + \beta}{2} H(i\omega)\right|^2 \frac{1}{2\pi} \int_{-\infty}^{\infty} |F(i\omega)|^2 \, d\omega$$

$$= \inf_{\omega \in (-\infty, \infty)} \left|1 + \frac{\alpha + \beta}{2} H(i\omega)\right|^2 \|f\|^2. \qquad \text{(III.4-18)}$$

Using (III.4-15) and Theorem I.4.2, we conclude that $(I + \tfrac{1}{2}(\alpha + \beta)L)$ has a bounded linear inverse with norm

$$\left\|\left(I + \frac{\alpha + \beta}{2} L\right)^{-1}\right\| \leq \sup_{\omega \in (-\infty, \infty)} \left|\left(1 + \frac{\alpha + \beta}{2} H(i\omega)\right)\right|^{-1}. \qquad \text{(III.4-19)}$$

In using Theorem I.4.2, it must be verified that $[I + \tfrac{1}{2}(\alpha + \beta)L]$ maps $L_2(-\infty, \infty)$ *onto* itself. This may be done by referring to the remark to the inversion formula (III.3-20) and considering $[1 + \tfrac{1}{2}(\alpha + \beta)H(i\omega)]^{-1}G(i\omega)$. Note that (III.4-5) as a condition for the existence of the inverse $(I + \tfrac{1}{2}(\alpha + \beta)L)^{-1}$ is much less restrictive then the use of Theorem I.4.3 which would require that

$$\sup_{\omega \in (-\infty,\,\infty)} \left| \frac{\alpha + \beta}{2} H(i\omega) \right| < 1. \qquad (\text{III.4-20})$$

The established existence of the inverse $(I + \frac{1}{2}(\alpha + \beta)L)^{-1}$ does not yet specify the form of the inverse or of $(I + \frac{1}{2}(\alpha + \beta)L)^{-1}L$ (can it be described by a convolution operation?). In the appendix to this chapter, it is shown that $g = (I + \frac{1}{2}(\alpha + \beta)L)^{-1}Lf$ may be described by

$$g(t) = \int_{-\infty}^{\infty} k(t - \tau) f(\tau)\, d\tau, \qquad (\text{III.4-21})$$

where k is an $L_1(-\infty, \infty)$ function which has as its Fourier transform, $[1 + \frac{1}{2}(\alpha + \beta)H(i\omega)]^{-1}H(i\omega)$. We relegate the proof to an appendix because it uses a somewhat advanced result. When $H(i\omega)$ is a rational function of $i\omega$, the proof is simpler (using partial fraction expansions). We do not, however, make this assumption concerning $H(i\omega)$ which would essentially confine us to integral equations derivable from differential or integro-differential equations (an important special case but less general than we wish to be).

Exercise III.4.2

In connection with the above paragraph, give the proof for the case when $H(i\omega)$ is a rational function of $i\omega$. Also, give a physical example leading to a nonrational $H(i\omega)$.

Condition (III.4-6) is now easily obtained from the expression for the linear operator norm (III.4-9) and (III.1-3) with $c = \frac{1}{2}(\alpha + \beta)$:

$$\|(I + cL)^{-1}L\| \eta(c) = \sup_{\omega \in (-\infty,\infty)} \left| \frac{H(i\omega)}{1 + \frac{1}{2}(\alpha + \beta)H(i\omega)} \right| \frac{1}{2}(\beta - \alpha) < 1.$$

$$(\text{III.4-22})$$

III.5 PREVUE OF STABILITY RESULTS

The circle condition of Sec. III.4 is strongly reminiscent of Nyquist's criterion for linear feedback systems. In particular, if $\alpha = \beta > 0$, then the circle condition of Sec. III.4, which is a sufficient condition for unique response to $L_2(-\infty, \infty)$ inputs, requires the locus of $H(i\omega)$ not to intersect the point $(-\beta^{-1}, 0)$ in the complex plane [see (III.4-5)]. On the other hand (loosely speaking), Nyquist's criterion for the stability of a linear feedback system with gain β and transfer function $H(i\omega)$ requires that the locus of $H(i\omega)$ not intersect the point $(-\beta^{-1}, 0)$ and, in addition, there is a stipulation about not encircling that point. We shall find that when we consider stability of nonlinear feedback systems in Chapter VIII (we postpone the definition of stability until that discussion) that we require the locus of $H(i\omega)$ not to encircle the circle when $\alpha > 0$ (in addition to not entering it). The nonencirclement is a consequence of requiring the operator $\left[I + \frac{1}{2}(\alpha + \beta)L \right]$ to be invertible in a different space than that used in Sec. III.4. More specifically, in this chapter, functions are defined on $(-\infty, \infty)$ while in Chapter VIII, functions are defined on $[0, \infty)$ so that initial conditions at $t = 0$ may be taken into account.

III.6 NOTES

Sections III.1 and III.2 are based on Sandberg [1]. A good basic reference on Fourier transforms is Goldberg. For convolution of an $L_2(-\infty, \infty)$ or $L_1(-\infty, \infty)$ function with an $L_2(-\infty, \infty)$ function, see Titchmarsh [1], p. 90 (Titchmarsh uses the terminology resultant for convolution). Convolution of an $L_1(-\infty, \infty)$ function with an $L_p(-\infty, \infty)$ function is discussed on pp. 99, 100 of Bochner and Chandrasekharan.

The approach to minimizing contraction constants given in this chapter serves our purpose well and is also relevant to the study of stability as we shall see in Chapter VIII. There has been other work on reducing the contraction constant for systems of the form of (III.1-1). See, e.g., Zames [1] and the book by Saaty for brief discussions of and references to the work by Kolodner and Zarantello. The reader is also referred to the chapter on Hammerstein equations by Dolph and Minty in Anselone.

Section III.4 and the Appendix are based on Sandberg [2].

In Section III.5, we mentioned Nyquist's criterion. The usual version of this criterion for rational functions can be found in many places, e.g., Kaplan, Chapter 7 or Zadeh and Desoer, Chapter 9. We state a version of the criterion in Sec. VI.4. Some recent work on generalizing the criterion is reported in Desoer and Wu.

APPENDIX

Reciprocal of Transforms; Fourier-Stieltjes

Transforms

To show that

$$\inf_{\omega \in (-\infty,\infty)} |1 + cH(i\omega)| > 0 \qquad \text{(III.A-1)}$$

implies that $(I + cL)^{-1}L$ can be described by the convolution operation (III. 4-21), we must back up and generalize the definition of Fourier transform. To motivate this generalization, first assume that $G(i\omega)$ is the Fourier transform of $g \in L_1(-\infty, \infty)$. Under what conditions is $[G(i\omega)]^{-1}$ the Fourier transform of an $L_1(-\infty, \infty)$ function? The answer is that there are no such conditions because the Riemann–Lebesgue lemma prevents $[G(i\omega)]^{-1}$ from being bounded, a necessary condition for the Fourier transform of an $L_1(-\infty, \infty)$ function [see (III.3-5)]. Hence, in order for that question to be meaningful, an extended definition of Fourier transform is required. The assumption (III.A-1) implies that $[1 + cH(i\omega)]$ is not the Fourier transform of an $L_1(-\infty, \infty)$ function and there is no reason to believe that $[1 + cH(i\omega)]^{-1}$ is. In fact, for $[1 + cH(i\omega)]^{-1}$ to be the Fourier transform of an $L_1(-\infty, \infty)$ function, we would require that $|H(i\omega)| \to \infty$ as $|\omega| \to \infty$, an impossibility.

Hence we are led to the *Fourier–Stieltjes transform* $F(i\omega)$ defined by

$$F(i\omega) = \int_{-\infty}^{\infty} e^{-i\omega t} \, df(t), \tag{III.A-2}$$

where f is a *function of bounded variation*. A property of (not the definition of†) a function of bounded variation f defined on $(-\infty, \infty)$ is that it can be broken up into three components:

$$f(t) = f_1(t) + f_2(t) + f_3(t), \tag{III.A-3}$$

where f_1, f_2, f_3 are the following kinds of functions:

f_1: absolutely continuous (which implies that it is differentiable almost everywhere); $\dot{f}_1 \in L_1(-\infty, \infty)$

f_2: constant except for a countable number of jumps

f_3: the singular part; contiuuous with zero derivative almost everywhere but it is not constant (which means that it is weird).

(III.A-2) may be decomposed as follows‡:

$$F(i\omega) = \int_{-\infty}^{\infty} e^{-i\omega t} \dot{f}_1(t) \, dt + \sum_n s_n e^{-i\omega t_n} + \int_{-\infty}^{\infty} e^{-i\omega t} \, df_3(t), \tag{III.A-4}$$

where s_n is the jump at $t = t_n$. If the last two terms in (III.A-4) are zero then the Fourier–Stieltjes transform is the Fourier transform of the $L_1(-\infty, \infty)$ function $\dot{f}_1$.

If $G(i\omega)$ is the Fourier transform of $g \in L_1(-\infty, \infty)$†† and $F(i\omega)$ is the Fourier–Stieltjes transform (III.A-2), then a convolution operation is defined by

$$k(t) = \int_{-\infty}^{\infty} g(t - \tau) \, df(\tau) \tag{III.A-5}$$

and $k \in L_1(-\infty, \infty)$. Furthermore,

$$K(i\omega) = G(i\omega)F(i\omega) \tag{III.A-6}$$

with $K(i\omega)$ the Fourier transform of k (Pitt, pp. 96, 97).

We need one more result: If $G(i\omega)$ is the Fourier transform of an $L_1(-\infty, \infty)$ function and

---

† For an introduction to functions of bounded variation and Stieltjes integrals (Riemann–Stieltjes integrals), see Natanson (he uses the terminology finite variation).

‡ Those familiar with distributions will recognize that f_2 introduces delta functions.

†† This g is not the same g as in (III. 4-21).

$$\inf_{\omega \in (-\infty, \infty)} |1 + G(i\omega)| > 0, \tag{III.A-7}$$

then $[1 + G(i\omega)]^{-1}$ is a Fourier–Stieltjes transform (Pitt, p. 99).

Now back to the original problem; assumption (III.A-1) implies that there is a function f of bounded variation such that

$$[1 + cH(i\omega)]^{-1} = \int_{-\infty}^{\infty} e^{-i\omega t}\, df(t). \tag{III.A-8}$$

Then,

$$k(t) = \int_{-\infty}^{\infty} h(t - \tau)\, df(\tau) \tag{III.A-9}$$

is in $L_1(-\infty, \infty)$ and has $H(i\omega)[1 + cH(i\omega)]^{-1}$ as its Fourier transform.

IV

Periodic Solutions–Global Contraction

This chapter considers the response of a nonlinear feedback system to a periodic input. Since the exact determination of the response is difficult or impossible, approximate techniques are used. The question of how close are the approximate and exact solutions naturally arises. We shall find that the analysis of the last chapter can be applied to answer this question for a class of systems. The condition for minimum contraction constant will again turn out to be a circle condition. We shall discuss the implications of violating this circle condition.

We first wish to discuss periodic functions and spaces of such functions (we need a Banach space to use the result of the last chapter). Some important properties of a particular space of periodic functions are shared in general by Hilbert spaces which we introduce. Our brief study of Hilbert spaces will shed some light on the nature and implications of some common approximation techniques.

IV.1 PERIODIC FUNCTIONS, FOURIER SERIES, AND HILBERT SPACES

The reader undoubtedly has had exposure to Fourier series. In this section, we state some basic facts concerning them.

A periodic function of period $T > 0$ (a *T-periodic function*) is one which is defined on $(-\infty, \infty)$ and which satisfies $x(t + T) = x(t)$ for all $t \,\epsilon(-\infty, \infty)$. The most familiar periodic functions are, of course, the sines and cosines. Assuming the following integrals exist, the Fourier coefficients of the real-valued T-periodic function x are

$$a_n = \frac{2}{T} \int_0^T x(t) \cos n\omega_0 t \, dt \quad n = 0, 1, 2, \ldots \qquad \text{(IV.1-1)}$$

$$b_n = \frac{2}{T} \int_0^T x(t) \sin n\omega_0 t \, dt \quad n = 1, 2, \ldots \qquad \text{(IV.1-2)}$$

with $\omega_0 = 2\pi/T$. The Fourier series of x is expressed as the right-hand side of (VI.1-3.)

$$x(t) \sim \frac{a_0}{2} + \sum_{n=1}^{\infty} (a_n \cos n\omega_0 t + b_n \sin n\omega_0 t). \qquad \text{(IV.1-3)}$$

The notation $\sim$ means that we associate with $x(t)$ the infinite series on the right-hand side without regard to whether this series converges. We replace the symbol $\sim$ by $=$ only when it is known that the infinite series does converge and that its sum is, in fact, $x(t)$.

A sufficient condition for replacing $\sim$ by $=$ (in a certain sense) is given in the following theorem. A function x is *piecewise smooth* over the interval $[0, T]$ if the interval can be divided into a finite number of subintervals and in the interior of each subinterval x has a continuous derivative.

Theorem IV.1.1

If x is periodic and piecewise smooth, then its Fourier series converges to $x(t)$, where $x(t)$ is continuous and to $\frac{1}{2}(x(t+) + x(t-))$ where $x(t)$ is discontinuous.

It is of interest to note that continuity of x is not sufficient to allow replacement of $\sim$ by $=$. In particular, there are continuous periodic functions whose Fourier series diverge at points.† We also mention as a point of interest (we

---

† However, the Fourier series of a continuous periodic function does converge almost everywhere; this was proved in 1966 by L. Carleson.

will not use this in the book) that for periodic functions whose Fourier series diverge, there are alternative methods of defining the sum of the series to result in convergence to the periodic function (Tolstov, Chapter 6).

The reader has no doubt been exposed to the complex form of Fourier series equivalent to (IV.1-3):

$$x(t) \sim \sum_{-\infty}^{\infty} c_n e^{in\omega_0 t} \tag{IV.1-4}$$

$$c_n = \frac{1}{T} \int_0^T x(t) e^{-in\omega_0 t} \, dt \tag{IV.1-5}$$

where the c_n are related to the a_n and b_n by

$$c_0 = \frac{a_0}{2}, \qquad c_n = \frac{a_n - ib_n}{2}, \qquad c_{-n} = \overline{c_n} \tag{IV.1-6}$$

(the overbar on numbers denotes complex conjugate). If x is complex-valued, then we can use (IV.1-4) as a Fourier series, but $c_{-n} = \overline{c_n}$ no longer holds.

An important class of periodic functions of a given period are those which are square integrable over the period: $\int_0^T x^2(t) \, dt$ exists and is finite. The very useful properties possessed by this class are shared by other classes of functions and most generally by Hilbert spaces which we introduce below. Hilbert spaces have properties which resemble those of a Euclidean space.

A linear space X is called an *inner product space* if, for any $x, y \in X$, there is defined an *inner product* $\langle x, y \rangle$ which is a number (complex or real depending on the field $\mathscr{F}$) satisfying

(i) $\langle y, x \rangle = \langle \overline{x, y} \rangle$ (IV.1-7)
(ii) $\langle \lambda x_1 + \mu x_2, y \rangle = \lambda \langle x_1, y \rangle + \mu \langle x_2, y \rangle$ $\lambda, \mu \in \mathscr{F}$ (IV.1-8)
(iii) $\langle x, x \rangle \geq 0;$ $\langle x, x \rangle = 0$ if and only if $x = 0.$ (IV.1-9)

If the space is real, (i) reads $\langle y, x \rangle = \langle x, y \rangle$.

The inner product is a generalization of the scalar product between two vectors in the plane, $\langle x, y \rangle = |xy| \cos \theta$, where θ is the angle between x and y. In fact, the more general inner product is also called a scalar product.

Exercise IV.1.1

Prove the following properties of inner products:

(a) $\langle x, \lambda y_1 + \mu y_2 \rangle = \overline{\lambda} \langle x, y_1 \rangle + \overline{\mu} \langle x, y_2 \rangle$
(b) $\langle x, 0 \rangle = \langle 0, y \rangle = 0.$

An important relation is the *Schwarz inequality* (also associated with the names of Cauchy and Bunyakovskii):

$$|\langle x, y \rangle|^2 \leq \langle x, x \rangle \langle y, y \rangle. \tag{IV.1-10}$$

To prove this, first note that if $y = 0$, then (IV.1-10) is obviously satisfied. Now, let $y \neq 0$ and consider

$$\langle x + \lambda y, x + \lambda y \rangle = \langle x, x \rangle + \bar{\lambda} \langle x, y \rangle + \lambda \langle y, x \rangle + |\lambda|^2 \langle y, y \rangle. \tag{IV.1-11}$$

Noting that (IV.1-11) must be non-negative for all λ [condition (iii)] and letting

$$\lambda = -\frac{\langle x, y \rangle}{\langle y, y \rangle}, \tag{IV.1-12}$$

we have

$$\langle x, x \rangle - \frac{|\langle x, y \rangle|^2}{\langle y, y \rangle} - \frac{|\langle x, y \rangle|^2}{\langle y, y \rangle} + \frac{|\langle x, y \rangle|^2}{\langle y, y \rangle} \geq 0 \tag{IV.1-13}$$

or

$$\langle x, x \rangle \langle y, y \rangle - |\langle x, y \rangle|^2 \geq 0. \tag{IV.1-14}$$

Any inner product space can become a normed linear space if we put

$$\|x\| = \sqrt{\langle x, x \rangle}. \tag{IV.1-15}$$

The first two conditions for a norm are immediately seen to be satisfied. The verification of the triangle inequality uses the Schwarz inequality:

$$\|x + y\|^2 = \langle x + y, x + y \rangle = |\langle x, x \rangle + \langle x, y \rangle + \langle y, x \rangle + \langle y, y \rangle|$$
$$\leq \|x\|^2 + 2\|x\| \cdot \|y\| + \|y\|^2 = (\|x\| + \|y\|)^2. \tag{IV.1-16}$$

With (IV.1-15), the following (parallelogram law) is easily verified:

$$\|x + y\|^2 + \|x - y\|^2 = 2[\|x\|^2 + \|y\|^2]. \tag{IV.1-17}$$

Whenever we speak of inner product spaces as normed linear spaces we assume that the norm and inner product are related by (IV.1-15). Similarly, all of our references to normed linear spaces as metric spaces assume that (I.3-1) is satisfied. Since an inner product space can be considered to be a normed linear space which is in turn a metric space, an inner product space may be complete. A complete inner product space is called a *Hilbert space*. Any Euclidean space R^n (the space of real n-vectors) is a Hilbert space with inner product

$$\langle x, y \rangle = \sum_{i=1}^{n} x_i y_i \qquad\qquad \text{(IV.1-18)}$$

with subscripts referring to components of vectors.

Another example of a Hilbert space is $L_2(a, b)$ introduced in Sec. III.3 if the inner product is

$$\langle x, y \rangle = \int_{a}^{b} x(t)y(t)\, dt. \qquad\qquad \text{(IV.1-19)}$$

Thus, the inequality in (III.3-2) is a special case of the general Schwarz inequality for inner product spaces [letting $x(t) = |f(t)|$, $y(t) = |g(t)|$]. None of the other $L_p(a, b)$ spaces are inner product spaces nor is $C(a, b)$.

The Hilbert space we shall use in this chapter is the space of real-valued T-periodic functions (T a fixed positive number) which are square-integrable over a period. The inner product is

$$\langle x, y \rangle = \frac{1}{T} \int_{0}^{T} x(t)y(t)\, dt. \qquad\qquad \text{(IV.1-20)}$$

Hence, the norm is the rms value, i.e.,

$$\|x\|^2 = \frac{1}{T} \int_{0}^{T} x^2(t)\, dt. \qquad\qquad \text{(IV.1-21)}$$

We shall call this space B_T.

The notion of orthogonality in Euclidean spaces is a familiar one. In the plane, two vectors are orthogonal when the angle between them is $90°$, i.e., the scalar product is zero. In an arbitrary Hilbert space H, two elements $x, y \in H$ are said to be *orthogonal* if $\langle x, y \rangle = 0$; orthogonality is denoted by $x \perp y$. If x is orthogonal to every $y \in A \subset H$, we write $x \perp A$. If A_1, $A_2 \subset H$ and $x_1 \perp x_2$ for all $x_1 \in A_1$, all $x_2 \in A_2$, then we write $A_1 \perp A_2$. The reader may find it useful to consider these definitions as applied to a Euclidean space, for example R^3 (a vector along one axis is orthogonal to a vector along another axis, an axis is orthogonal to the perpendicular plane, etc.).

Some simple properties of orthogonality are obvious. For example, if $x \perp y_1$ and $x \perp y_2$ then $x \perp \lambda_1 y_1 + \lambda_2 y_2$ with $\lambda_1, \lambda_2 \in \mathscr{F}$. An important theorem is concerned with the *orthogonal complement* of a set $A \subset H$, which is the set of all elements orthogonal to A. The orthogonal complement may be shown to be a closed subspace of H. (Subspaces are defined at the end of Sec. I.3).

The theorem to be given is a far-reaching generalization of the observation that the point x on a line closest to a point y not on the line is obtained by dropping a perpendicular from y to the line. Note that $(x - y)$ is orthogonal to the line. The following theorem generalizes the line to a subspace of a Hilbert space. (We state without proof the rest of the theorems in this section.)

Theorem IV.1.2

Let H_1 be a closed subspace of a Hilbert space H and let H_2 be its orthogonal complement. Then each $x \in H$ can be expressed uniquely† as

$$x = x_1 + x_2 \qquad x_1 \in H_1, \qquad x_2 \in H_2. \qquad \text{(IV.1-22)}$$

Furthermore

$$\|x - x_1\| = d(x, H_1), \qquad \text{(IV.1-23)}$$

i.e., x_1 is the element of H_1 closest to x. [By distance of a point x to a set A, we mean inf $d(x, y)$ where d is the metric.]
$\quad y \in A$

Referring to Theorem IV.1.2, the linear mappings P_1 and P_2, defined by $x_1 = P_1 x$ and $x_2 = P_2 x$, are called *projections*. Projections can be put in a more general context. Suppose that a linear space X can be decomposed as a *direct sum*

$$X = X_1 + X_2 + \cdots + X_n \qquad \text{(IV.1-24)}$$

where the X_i are subspaces of X and each $x \in X$ has a unique representation of the form

$$x = x_1 + x_2 + \cdots + x_n \qquad \text{(IV.1-25)}$$

with $x_i \in X_i$. Then the mappings P_i defined by $x_i = P_i x$ are projections. The notion of projection is familiar in, say, the plane where we can project a vector onto the horizontal and vertical axes.

Exercise IV.1.2

Prove:

(a) $X_i \cap X_j = \{0\}$ $i \neq j$ (0 is zero element)
(b) $P_i P_j = 0$ $i \neq j$ (0 is null operator, i.e., it maps into zero element)
(c) $P_i^2 = P_i$ $i = 1, \ldots, n$
(d) $\sum_{i=1}^{n} P_i = I$ (the identity operator)
(e) in Hilbert spaces, $\|P_i\| = 1$ (X_i assumed not to be the trivial subspace $\{0\}$)

† We do not consider the addition of a zero element to affect uniqueness.

   Theorem IV.1.2 is a concrete example of a direct sum decomposition for Hilbert spaces. In that case, $X_1 = H_1$ and $X_2 = H_2$ are orthogonal and the mappings defined by $x_1 = P_1 x$ and $x_2 = P_2 x$ are called *orthogonal projections*. This result for Hilbert spaces is one reason they are convenient to work with. It is not always possible to conveniently decompose normed linear spaces.

   One great advantage of working with a Hilbert space is that one often can express an arbitrary member of the Hilbert space as a linear combination of some elementary elements in the space. In the plane, any vector can be represented as the sum of two vectors perpendicular to each other. In this chapter, our specific interest is in expressing periodic functions (square integrable over a period) as a sum of sines and cosines or complex exponentials (a Fourier series). We thus wish to delineate certain subsets of Hilbert spaces. We shall refer to such subsets as systems $\{x_a\}$ with $a \in A$, an indexing set. We make the following definitions for a system $\{x_a\} \subset H$:

(i) $\mathscr{L}(\{x_a\})$ is the set of all elements of the form $x = \lambda_1 x_{a1} + \cdots + \lambda_n x_{an}$ where $x_{a1}, \ldots, x_{an} \in \{x_a\}$, and $\lambda_1, \ldots, \lambda_n \in \mathscr{F}$. $\mathscr{L}(\{x_a\})$ is called the subspace spanned by $\{x_a\}$.

(ii) $\{x_a\}$ is *complete* if $\overline{\mathscr{L}(\{x_a\})} = H$ (the overbar on a set denotes closure).

(iii) $\{x_a\}$ is *orthogonal* if, for any $x_{ai}, x_{aj} \in \{x_a\}$, we have $x_{ai} \perp x_{aj}$ if $x_{ai} \neq x_{aj}$.

(iv) $\{x_a\}$ is *orthonormal* if it is orthogonal and also $\|x_a\| = 1$, all $a \in A$.

   The definition of completeness for a system of elements is distinct from completeness of a metric space. The context will always make it clear which usage is meant. A system is complete if we take the set of finite linear combinations of members of the system and the closure of that set is the Hilbert space. In other words, any element in the Hilbert space can be approximated arbitrarily closely by a finite linear combination of elements of the system. To convert an orthogonal system to an orthonormal system, one need merely divide each (nonzero) element by its norm. An example of a complete orthonormal system in B_T is the system of trigonometric functions

$$\{1, \sqrt{2}\ \cos \omega_0 t, \sqrt{2}\ \sin \omega_0 t, \sqrt{2}\ \cos 2\omega_0 t, \ldots \}. \qquad \text{(IV.1-26)}$$

The system of complex exponentials $\{e^{in\omega_0 t}\}$ is not a complete system for B_T because the functions in B_T are real while the $e^{in\omega_0 t}$ are complex. However, we may consider the complex exponentials as a complete orthonormal system for B_T if we take linear combinations with complex multipliers c_n which satisfy $c_{-n} = \overline{c_n}$ (and using the resulting equivalence with the sines and cosines). We shall find this to be a notational convenience in the next section.

   If the Hilbert space satisfies a condition of separability, then its orthonormal systems have very useful properties. A metric space X is *separable* if there is a countable set $D \subset X$ such that its closure $\overline{D} = X$. We shall not elaborate on separability here but merely state that the Hilbert spaces we shall use (including B_T) are separable and henceforth when we say Hilbert space, we mean separable Hilbert space.

Theorem IV.1.3

An orthonormal system $\{x_a\}(a \in A)$ in a (separable) Hilbert space is countable.

Theorem IV.1.4

There exists in every (separable) Hilbert space a countable complete orthonormal system.

We have already listed a countable complete orthonormal system for B_T. In general, let $\{x_k\}(k = 1, 2, \dots)$ be an orthonormal system in the separable Hilbert space H. Note that in view of Theorem IV.1.3, the indexing set can be taken as that set of positive integers (or all integers). If $x \in H$, the numbers

$$a_k = \langle x, x_k \rangle \qquad k = 1, 2, \dots \tag{IV.1-27}$$

are called the *Fourier coefficients* of x with respect to the given orthonormal system. The *Fourier series* of x is defined to be

$$\sum_{k=1}^{\infty} a_k x_k = \sum_{k=1}^{\infty} \langle x, x_k \rangle x_k. \tag{IV.1-28}$$

These definitions reduce to the familiar ones given at the beginning of this section if the trigonometric system (IV.1-26) in B_T is considered except for scaling which merely depends on the norm (and could be changed by rescaling the norm). From (IV.1-26) and (IV.1-27), we have for B_T (with relabeling),

$$a_0 = \frac{1}{T} \int_0^T x(t)\, dt \tag{IV.1-29}$$

$$a_k = \frac{\sqrt{2}}{T} \int_0^T x(t) \cos k\omega_0 t\, dt \qquad k = 1, 2, \dots \tag{IV.1-30}$$

$$b_k = \frac{\sqrt{2}}{T} \int_0^T x(t) \sin k\omega_0 t\, dt \qquad k = 1, 2, \dots. \tag{IV.1-31}$$

An important property of the Fourier series of x is that the first n elements of the series are a best approximation to x when compared with any other linear combination of the first n of the x_k's. To state this precisely, let the partial sum of the Fourier series be denoted s_n,

$$s_n(x) = \sum_{k=1}^{n} a_k x_k, \tag{IV.1-32}$$

where the a_k are the Fourier coefficients.

Theorem IV.1.5

The partial sum $s_n(x)$ of the Fourier series x results from projection of x onto $\mathscr{L}(\{x_1, x_2, \ldots, x_n\})$, the subspace spanned by $\{x_1, x_2, \ldots, x_n\}$.

Theorem IV.1.5, together with Theorem IV.1.2, gives the next important theorem concerning the best approximation property of Fouries series mentioned before Theorem IV.1.5.

Theorem IV.1.6

For any $b_1, \ldots, b_n \in \mathscr{F}$,

$$\|x - s_n(x)\| \le \left\| x - \sum_{k=1}^{n} b_k x_k \right\|.$$

Though we now have a most important result concerning partial sums of Fourier series we have not yet said anything about convergence of the series.

Theorem IV.1.7

If the system $\{x_k\}$ is complete, the Fourier series of any $x \in H$ is convergent to x.

Theorem IV.1.7 states that, with a complete system,

$$\lim_{n \to \infty} s_n(x) = \lim_{n \to \infty} \sum_{k=1}^{n} \langle x, x_k \rangle x_k = x. \tag{IV.1-33}$$

With a complete system, we also have

$$\|x\|^2 = \sum_{k=1}^{\infty} |a_k|^2 = \sum_{k=1}^{\infty} |\langle x, x_k \rangle|^2 \tag{IV.1-34}$$

which is often useful in evaluating norms of elements. Relation (IV.1-34) is called *Parseval's equality*.

The following theorem is concerned with the question of when a given sequence of numbers are the Fourier coefficients of an element in a Hilbert space.

Theorem IV.1.8 (*Riesz-Fischer*)

Given a numerical sequence $\{c_k\}$ for which $\sum\limits_{k=1}^{\infty} |c_k|^2 < \infty$ and a complete orthonormal system in a Hilbert space H, there exists a unique $x \in H$ such that all its Fourier coefficients $a_k = c_k$.

IV.2 SYSTEM TO BE STUDIED

We shall consider the special case of the system studied in Chapter III (see (III.1-1) and Fig. III.1) when the space is B_T. Hence the system is described by

$$x = -LNx + r, \ r \in B_T. \tag{IV.2-1}$$

To guarantee that N maps B_T into itself, $y = Nx$ is defined by

$$y(t) = n[x(t)] \tag{IV.2-2}$$

with the real-valued function n satisfying the slope condition: There are two real constants α and β ($\beta > 0$) such that

$$\alpha \le \frac{n(u_1) - n(u_2)}{u_1 - u_2} \le \beta \tag{IV.2-3}$$

for all real $u_1 \ne u_2$. It is easily verified that N maps B_T into itself. To specify L, we use the complete orthonormal system for B_T consisting of sines and cosines and represent them by the complex exponentials $\{e^{ik\omega_0 t}\}$ with the proviso that $c_{-k} = \overline{c_k}$. If $x \in B_T$, then

$$\begin{aligned} x(t) &= \sum_{k=-\infty}^{\infty} c_k e^{ik\omega_0 t} \\ c_k &= \frac{1}{T} \int_0^T x(t) e^{-ik\omega_0 t} \, dt. \end{aligned} \tag{IV.2-4}$$

Now we define the operator L by means of the countable set of complex constants $\{\ldots, H_{-2}, H_{-1}, H_0, H_1, H_2, \ldots\}$ satisfying $\sup\limits_{k} |H_k| < \infty$ and $H_k = \overline{H_{-k}}$. If $y = Lx$ and x is given by (IV.2-4), then

$$y(t) = \sum_{k=-\infty}^{\infty} b_k e^{ik\omega_0 t}, \tag{IV.2-5}$$

where

$$b_k = H_k c_k. \tag{IV.2-6}$$

That $Lx \in B_T$ if $x \in B_T$ follows from

$$\sum_{k=-\infty}^{\infty} |b_k|^2 \leq \sup_k |H_k|^2 \sum_{k=-\infty}^{\infty} |c_k|^2 < \infty \qquad \text{(IV.2-7)}$$

and the Riesz–Fischer theorem (Theorem IV.1.8). The constants H_k are typically the values of a transfer function $H(i\omega)$ evaluated at $\omega = k\omega_0$. More precisely, the linear operator L might arise from the convolution,

$$y(t) = \int_{-\infty}^{\infty} h(t - \tau)x(\tau)\, d\tau \qquad \text{(IV.2-8)}$$

with h assumed to be a function in $L_1(-\infty, \infty)$. To show that $y \in B_T$ if $h \in L_1(-\infty, \infty)$ and $x \in B_T$, we first use the Schwarz inequality as follows:

$$|y(t)| \leq \int_{-\infty}^{\infty} |x(t - \tau)| \cdot |h(\tau)|^{1/2} \cdot |h(\tau)|^{1/2}\, d\tau$$

$$\leq \left(\int_{-\infty}^{\infty} |x(t - \tau)|^2 \cdot |h(\tau)|\, d\tau \right)^{1/2} \left(\int_{-\infty}^{\infty} |h(u)|\, du \right)^{1/2}. \qquad \text{(IV.2-9)}$$

Concerning the use of the Schwarz inequality, note that $h \in L_1(-\infty, \infty)$ implies $|h|^{1/2} \in L_2(-\infty, \infty)$, but we do not know that $|x(t - \tau)| \cdot |h(\tau)|^{1/2}$ is in $L_2(-\infty, \infty)$. If it is not, then (IV.2-9) is interpreted as $|y(t)| \leq \infty$ which is not too informative but true. Then,

$$\int_0^T |y(t)|^2\, dt \leq \int_0^T \left[\int_{-\infty}^{\infty} |x(t - \tau)|^2 \cdot |h(\tau)|\, d\tau \right] dt \int_{-\infty}^{\infty} |h(u)|\, du. \qquad \text{(IV.2-10)}$$

It may be shown that the orders of integration can be interchanged† so that

$$\int_0^T |y(t)|^2\, dt \leq \int_0^T |x(t)|^2\, dt \left(\int_{-\infty}^{\infty} |h(\tau)|\, d\tau \right)^2. \qquad \text{(IV.2-11)}$$

That $y(t)$ is real-valued and periodic with period T is clear. Hence, $y \in B_T$.

To show how H_k arises from $H(i\omega)$, we have (using a justifiable change of order of integration)

$$\int_0^T y(t)e^{-ik\omega_0 t}\, dt = \int_0^T \left[\int_{-\infty}^{\infty} h(\tau)x(t - \tau)\, d\tau \right] e^{-ik\omega_0 t}\, dt$$

$$= \int_{-\infty}^{\infty} \left[\int_0^T x(t - \tau)e^{-ik\omega_0 t}\, dt \right] h(\tau)\, d\tau$$

$$= \int_{-\infty}^{\infty} \left[\int_0^T x(t - \tau)e^{-ik\omega_0(t - \tau)}\, dt \right] h(\tau)e^{-ik\omega_0 \tau}\, d\tau$$

$$= H(ik\omega_0) \int_0^T x(t)e^{-ik\omega_0 t}\, dt. \qquad \text{(IV.2-12)}$$

† For those familiar with Fubini's theorem: using the periodicity of x,

$$\int_{-\infty}^{\infty} \left[\int_0^T |x(t - \tau)|^2\, dt \right] |h(\tau)|\, d\tau = \int_0^T |x(t)|^2\, dt \int_{-\infty}^{\infty} |h(\tau)|\, d\tau < \infty.$$

We noted that we could not conclude in the above that y is bounded. The following additional condition on h will guarantee boundedness:

$$\int_{-\infty}^{\infty} (1 + t^2)[h(t)]^2 \, dt < \infty. \tag{IV.2-13}$$

First observe that with $x \in B_T$,

$$\int_{-\infty}^{\infty} \frac{[x(t-\tau)]^2}{1+\tau^2} \, dt = \sum_{n=-\infty}^{\infty} \int_{nT}^{(n+1)T} \frac{[x(t-\tau)]^2}{1+\tau^2} \, d\tau$$

$$\leq 2\left(1 + T^{-2} \sum_{n=1}^{\infty} n^{-2}\right) \int_0^T [x(t)]^2 \, dt < \infty. \tag{IV.2-14}$$

Hence, using the Schwarz inequality (IV.1-10) for the space $L_2(-\infty, \infty)$,

$$|y(t)|^2 = \left| \int_{-\infty}^{\infty} (1 + \tau^2)^{1/2} h(\tau) \frac{x(t-\tau)}{(1+\tau^2)^{1/2}} \, d\tau \right|^2$$

$$\leq \int_{-\infty}^{\infty} (1 + \tau^2)[h(\tau)]^2 \, d\tau \int_{-\infty}^{\infty} \frac{[x(t-\tau)]^2}{1+\tau^2} \, d\tau \tag{IV.2-15}$$

proving that y is bounded.

We remark in passing that condition (IV.2-13) applied to the convolution operation of (IV.2-8) has an even stronger implication than mapping B_T into itself. If the real-valued measurable function x is such that

$$\lim_{T \to \infty} \frac{1}{2T} \int_{-T}^{T} |x(t)|^2 \, dt < \infty, \tag{IV.2-16}$$

then $y(t)$ given by (IV.2-8) also satisfies (IV.2-16). This follows from the result that (IV.2-16) implies that

$$\int_{-\infty}^{\infty} \frac{[x(t)]^2}{1+t^2} \, dt < \infty \tag{IV.2-17}$$

and an argument similar to that used in deriving (IV.2-15) (details are given in Beneš). The class of functions satisfying (IV.2-16) includes all those in B_T but is much broader. For example, if x is an *almost-periodic* function of the form

$$x(t) = \sum_{m=1}^{n} a_m e^{i\lambda_m t} \tag{IV.2-18}$$

where the λ_m are not necessarily integral multiples of some fundamental frequency, then x satisfies (IV.2-16).

IV.3 CONTRACTION MAPPING ANALYSIS

The analysis of Sec. III.2 is immediately applicable to the system under consideration in this chapter. So under the same assumptions as given in Sec. III.1 (specialized to the space B_T), we form the system equivalent to (IV.2-1):

$$x = -(I + cL)^{-1}L(N - cI)x + (I + cL)^{-1}r, \quad c = \tfrac{1}{2}(\alpha + \beta).$$
$$\text{(IV.3-1)}$$

The condition for minimum contraction constant from Sec. III.2 is

$$\|[I + \tfrac{1}{2}(\alpha + \beta)L]^{-1}L\|\tfrac{1}{2}(\beta - \alpha) \leq \gamma < 1. \qquad \text{(IV.3-2)}$$

If successive approximations are used,

$$x_{m+1} = -(I + cL)^{-1}L(N - cI)x_m + (I + cL)^{-1}r \quad m = 0, 1, \dots$$
$$\text{(IV.3-3)}$$

and if (IV.3-2) is satisfied, then the fixed point x^* satisfies

$$\|x^* - x_m\| \leq \frac{\gamma^m}{1 - \gamma}\|x_1 - x_0\|$$

$$= \frac{\gamma^m}{1 - \gamma}\|(I + cL)^{-1}[-LNx_0 + r - x_0]\| \qquad \text{(IV.3-4)}$$

using an easily verified expression for $(x_1 - x_0)$. When $m = 0$, the condition (IV.3-4) gives a bound on the difference between the exact solution and any $x_0 \in B_T$. If x_0 is regarded as an approximate solution, we have a bound on the error of approximation.

It remains to specify how the assumptions made in Chapter 2 and the contraction condition (IV.3-2) are to be interpreted in B_T. Condition (IV.2-7) shows that L is a bounded linear operator of B_T into itself. From (I.4-5), (IV.2-7), and (IV.1-34), it follows that

$$\|L\| \leq \sup_k |H_k|. \qquad \text{(IV.3-5)}$$

That the inequality of (IV.3-5) may be replaced by equality follows by considering $y = Lx$ with $|H_{k'}| = \sup_k |H_k|$ and

$$x(t) = \frac{e^{ik'\omega_0 t} + e^{-ik'\omega_0 t}}{\sqrt{2}}. \qquad \text{(IV.3-6)}$$

Since $\|x\| = 1$, we have

$$\|y\| = |H_{k'}| = \sup_k |H_k|. \tag{IV.3-7}$$

Then from (IV.3-5) and (IV.3-7), we obtain

$$\|L\| = \sup_k |H_k|. \tag{IV.3-8}$$

The reader should have caught a slight defect in the above argument, namely that for some specific k', $|H_{k'}| = \sup_k |H_k|$. We cannot assume that the supremum is actually achieved.† But from the definition of sup, for any $\epsilon > 0$ there is a k' such that

$$|H_{k'}| > \sup_k |H_k| - \epsilon. \tag{IV.3-9}$$

Then it follows that for any $\epsilon > 0$ and some x with $\|x\| = 1$

$$\|Lx\| \geq \sup_k |H_k| - \epsilon. \tag{IV.3-10}$$

Now (IV.3-8) follows.

Next we discuss the invertibility of $\left[I + \frac{1}{2}(\alpha + \beta)L\right]$. A sufficient condition for $\left[I + \frac{1}{2}(\alpha + \beta)L\right]$ to have a bounded linear inverse is that

$$\inf_k |1 + \tfrac{1}{2}(\alpha + \beta)H_k| = d > 0. \tag{IV.3-11}$$

To show this, let

$$x(t) = \sum_{k=-\infty}^{\infty} a_k e^{ik\omega_0 t} \in B_T. \tag{IV.3-12}$$

Then $y = \left[I + \frac{1}{2}(\alpha + \beta)L\right]x$ satisfies

$$\|y\|^2 \geq d^2 \sum_{k=-\infty}^{\infty} |a_k|^2 = d^2 \|x\|^2. \tag{IV.3-13}$$

The reader can verify that $\left[I + \frac{1}{2}(\alpha + \beta)L\right]$ is onto, so that by Theorem I.4.2 it has a bounded inverse. The inverse is clearly defined by an equation of the form of (IV.2-6) with H_k replaced by $\left[1 + \frac{1}{2}(\alpha + \beta)H_k\right]^{-1}$.

Finally, we need the norm of $\left[I + \frac{1}{2}(\alpha + \beta)L\right]^{-1}L$. Using the same arguments as those for $\|L\|$ we find

† We are here only assuming $\sup_k |H_k| < \infty$, not that H_k is necessarily derivable from an $L_1(-\infty, \infty)$ Fourier transform as in the discussion of (IV.2-8).

$$\|[I + \tfrac{1}{2}(\alpha + \beta)L]^{-1}L\| = \sup_k \left| \frac{H_k}{1 + \tfrac{1}{2}(\alpha + \beta)H_k} \right|. \tag{IV.3-14}$$

Then the condition for a contraction (IV.3-2) becomes

$$\sup_k \left| \frac{H_k}{1 + \tfrac{1}{2}(\alpha + \beta)H_k} \right| \tfrac{1}{2}(\beta - \alpha) < 1. \tag{IV.3-15}$$

Recapitulating, if the system described in Sec. IV.2 satisfies

$$\alpha \le \frac{n(u_1) - n(u_2)}{u_1 - u_2} \le \beta, \qquad u_1 \ne u_2$$

$$\sup_k |H_k| < \infty$$

$$\inf_k |1 + \tfrac{1}{2}(\alpha + \beta)H_k| > 0 \tag{IV.3-16}$$

$$\sup_k \left| \frac{H_k}{1 + \tfrac{1}{2}(\alpha + \beta)H_k} \right| \tfrac{1}{2}(\beta - \alpha) = \gamma < 1,$$

then for all inputs $r \in B_T$ there is a unique response $x^* \in B_T$ satisfying

$$x^* = -LN(x^*) + r. \tag{IV.3-17}$$

Furthermore, x^* satisfies (IV.3-4).

Several interesting observations can be made concerning this result. First of all, the conditions disallow the existence of a nontrivial self-sustained oscillation of period T (i.e., $x \ne 0$ for $r = 0$) if $n(0) = 0$ because the zero solution is then the unique solution. When H_k is the value of a transform $H(i\omega)$ evaluated at $\omega = k\omega_0$ as discussed at the end of Sec. IV.2,

$$\sup_{\omega \in (-\infty, \infty)} \left| \frac{H(i\omega)}{1 + \tfrac{1}{2}(\alpha + \beta)H(i\omega)} \right| \tfrac{1}{2}(\beta - \alpha) < 1 \tag{IV.3-18}$$

and $n(0) = 0$ disallows a nontrivial self-sustained oscillation of any period. Condition (IV.3-18) is furthermore a sufficient condition for the absence of jump resonance (nonunique response to periodic input) because of the uniqueness of the fixed points. We will discuss jump resonance in some detail in Sec. IV.5.

Let us note that (IV.3-18) is the same as (III.4-6) and has the same geometrical interpretation given below (III.4-6) with regard to the locus of $H(i\omega)$. For the purposes of Sec. IV.5, it will be slightly more convenient to give (IV.3-18) an interpretation with $H^{-1}(i\omega)$ rather than $H(i\omega)$. Condition (IV.3-18) is rewritten as

$$\inf_{\omega \in (-\infty, \infty)} |H^{-1}(i\omega) + \tfrac{1}{2}(\alpha + \beta)| > \tfrac{1}{2}\frac{(\beta - \alpha)}{2} \tag{IV.3-19}$$

which says the locus of $H^{-1}(i\omega)$ is outside and bounded away from the disc centered in the complex plane at $\left(-\frac{1}{2}(\alpha + \beta), 0\right)$ and with radius $\frac{1}{2}(\beta - \alpha)$ (see Fig. IV.1)

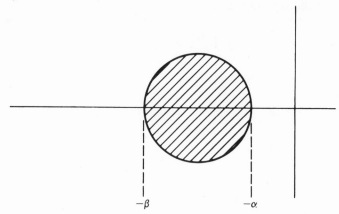

Figure IV.1 Forbidden region for $H^{-1}(i\omega)$

Similarly, condition (IV.3-15) states that the numbers $(H_k)^{-1}$ are bounded away from the disk of Fig. IV.1. Since the contraction mapping analysis only provides a sufficient condition, one may wonder how close to necessary it is. Well, if there is a point $(H_k)^{-1}$ which lies on the real axis diameter of the disk of Fig. IV.1, there will be self-sustained oscillations if the function n is linear with slope $(-H_k)^{-1}$ with $\alpha \leq (-H_k)^{-1} \leq \beta$, since $x(t) = \overline{a_k} e^{-ik\omega_0 t} + a_k e^{ik\omega_0 t}$ satisfies $x = -LNx + r$ for any a_k if $r = 0$. Pursuing this example further, if $r(t) = \overline{b_k} e^{-ik\omega_0 t} + b_k e^{ik\omega_0 t}$, $b_k \neq 0$, then there is no $x \in B_T$ satisfying $x = -LNx + r$. Hence, having the numbers $(H_k)^{-1}$ not touch the real axis diameter of the disc represents a necessary and sufficient condition for the existence of unique solutions in B_T for all inputs in B_T and all functions n satisfying the slope conditions. Section IV.5 further discusses implications of entering the disk. Let us just note here that with (IV.3-18) violated by the locus of $H(i\omega)$ cutting the real axis of the disc corresponds to violating Nyquist's criterion for a linear system satisfying the slope condition (IV.2-3).

IV.4 EQUIVALENT LINEARIZATION, DESCRIBING FUNCTIONS, HARMONIC BALANCE

The bound of (IV.3-4) depends on $\|x_1 - x_0\|$ which is what was denoted k in Sec. II.4. In the last section the contraction constant γ does not depend on x_0 so that picking x_0 to minimize k directly minimizes the bound (IV.3-4). This viewpoint will lead to suggesting use of some approximate techniques of solving (IV.2-1).

With this background, let us try to pick some approximate solution x_0 of (IV.2-1). Anticipating a contraction mapping analysis we wish to minimize $\|-LN(x_0) + r - x_0\|$, which will minimize a bound on k. Since the minimum is obviously zero when $x_0 = x^*$ and we presumably cannot exactly solve for x^*, we do the minimizing over some class of x_0's. It is appealing to let x_0 be the solution of an equation resembling (IV.2-1), but which is simpler to solve. To motivate this, first let

$$r(t) = \overline{a_1} e^{-i\omega_0 t} + a_1 e^{i\omega_0 t}. \tag{IV.4-1}$$

(This assumption is not necessary; it is only made for simplicity.) It thus seems reasonable to choose for a first approximation an x_0 of the form

$$x_0(t) = \overline{b_1} e^{-i\omega_0 t} + b_1 e^{i\omega_0 t} \tag{IV.4-2}$$

with x_0 as a solution (the existence of which is discussed later in this section) of the system

$$x_0 = -L k_e x_0 + r, \tag{IV.4-3}$$

where k_e is a constant, called the equivalent gain. Then,

$$\|-LN x_0 + r - x_0\| \le \|L\| \cdot \|N x_0 - k_e x_0\|. \tag{IV.4-4}$$

If

$$(N x_0)(t) = \sum_{k=-\infty}^{\infty} c_k e^{ik\omega_0 t}, \qquad c_0 = 0, \tag{IV.4-5}$$

then Theorem IV.1.6 says that $\|N(x_0) - k_e x_0\|$ is minimized when

$$k_e = \frac{c_1}{b_1} \left(= \frac{\bar{c}_1}{\bar{b}_1} \right). \tag{IV.4-6}$$

However, note that b_1 depends on x_0 which in turn depends on k_e.[†] Thus (IV.4-3) and (IV.4-6) must be simultaneously satisfied and (IV.4-3) is actually not a linear system because of the dependence of k_e on x_0. Nevertheless, (IV.4-3) is "almost linear" and it is much easier to work with than (IV.2-1).

Observe that k_e is the ratio of the first Fourier coefficients of the output and input of the nonlinearily. In control systems, k_e is called a describing function

---

† Hence, strictly speaking, k_e minimizes $\|N(x_0) - k_e x_0\|$ ignoring the dependence of x_0 on k_e.

and (IV.4-3) is often solved graphically especially when $r = 0$ (see e.g., Hsu and Meyer). In control systems literature, $x(t)$ is often assumed to equal $E \sin \omega t$ and the real form of the Fourier series is used. Then $k_e = (B_1 + iA_1)/E$ where A_1 and B_1 are the $\cos \omega_0 t$ and $\sin \omega_0 t$ Fourier coefficients of $(Nx_0)(t)$, respectively. A typical notation for the k_e is $N(E)$.†

Let us note that, while k_e minimizes $\|L\| \cdot \|Nx_0 - k_e x_0\|$, it does not necessarily minimize $\|(I + cL)^{-1} L(Nx_0 - k_e x_0)\|$ or even $\|LNx_0 - Lk_e x_0\|$. This latter minimization is much more complicated although some authors have done this. The motivation for describing functions used by control engineers is based on the smallness of $L[N(x_0) - k_e x_0]$ which represents the linear operator mapping the harmonics of $N(x_0)$ since $k_e x_0$ is the fundamental of $N(x_0)$ (assume for the moment that the constant terms in the Fourier series are zero). The control engineer usually assumes that the transfer function $H(i\omega)$ associated with L is low pass so that $H(ik\omega)$ is small for $k > 1$ and then $\|L[N(x_0) - k_e x_0]\|$ will tend to be small.

Recognizing that $Lk_e x_0$ is L operating on the fundamental of $N(x_0)$, we obtain another representation for x_0,

$$x_0 = -\tilde{L}N(x_0) + r, \tag{IV.4-7}$$

where $\tilde{L}$ is defined as follows: If

$$x(t) = \sum_{k=-\infty}^{\infty} a_k e^{ik\omega_0 t} \tag{IV.4-8}$$

(assume $a_0 = 0$), then

$$(\tilde{L}x)(t) = \overline{H_1 a_1} e^{-i\omega_0 t} + H_1 a_1 e^{i\omega_0 t}. \tag{IV.4-9}$$

$\tilde{L}$ is thus a linear operator which operates with L and then extracts the fundamental. This suggests a more general method of approximation than equivalent linearization. If x is given by (IV.4-8), let P_n map x into the partial sum of its Fourier series, i.e.,

$$(P_n x)(t) = \sum_{k=-n}^{n} a_k e^{ik\omega_0 t}. \tag{IV.4-10}$$

† $N(E)$ is real since

$$\pi A_1 = \int_0^{2\pi} n(E \sin \theta) \cos \theta \, d\theta$$
$$= \left[\int_0^{\pi/2} + \int_{\pi/2}^{\pi} + \int_{\pi}^{3\pi/2} + \int_{3\pi/2}^{2\pi} \right] n(E \sin \theta) \cos \theta \, d\theta = 0 ;$$

the first two and last two integrals cancel. In a similar manner, the k_e of (IV.4-6) is shown to be real. Note that the realness of $N(E)$ depends on the single-valuedness of n which we assume. The describing function of nonlinearities such as hysteresis can be complex. See Hsu and Meyer, p. 208.

The operator P_n acts like an ideal filter, rejecting all harmonics above a certain frequency. In view of Theorems IV.1.2 and IV.1.5, P_n is an orthogonal projection. When the averages of the periodic functions are zero, as is often the case, we have $P_1 L = \tilde{L}$ [see (IV.4-9) and (IV.4-10)]. An approximation method more general than (IV.4-7) is to let

$$x_0 = -P_n LN(x_0) + r. \tag{IV.4-11}$$

Observing that with (IV.4-11),

$$\|-LN(x_0) + r - x_0\| = \|(I - P_n)LN(x_0)\|, \tag{IV.4-12}$$

and, by referring to Theorem IV.1.6, we see that a bound on k is minimized by using P_n. Equation (IV.4-11) represents the method of harmonic balance and a version of Galerkin's method and also the dual input describing function method for $n = 2$. Equation (IV.4-11) is solved by balancing the Fourier coefficients on both sides of the equation.

It is easily verified that the hypotheses of the last section [see (IV.3-16)] also imply the existence of a unique solution in B_T to the approximate equation (IV.4-11) for any $n \geq 1$. For example,

$$\|[I + \tfrac{1}{2}(\alpha + \beta)P_n L]^{-1} P_n L\| = \max_{|k| \leq n} \left| \frac{H_k}{1 + \tfrac{1}{2}(\alpha + \beta)H_k} \right|$$
$$\leq \sup \left| \frac{H_k}{1 + \tfrac{1}{2}(\alpha + \beta)H_k} \right| \tag{IV.4-13}$$

so that conditions for a contraction are satisfied.

We close this section with an emphasis that the discussion has assumed use of the B_T norm of (IV.1-21), the rms norm. In the next chapter we shall use a Banach space of continuous periodic functions C_T with norm

$$\|x - y\|_c = \max_t |x(t) - y(t)|. \tag{IV.4-14}$$

With this norm, Theorem IV.1.6 is not generally true. The best trigonometric polynomial approximation to a periodic function is not given by the partial sum of its Fourier series. A *trigonometric polynomial of degree n* is a linear combination of functions in the set $\{1, \cos \omega_0 t, \sin \omega_0 t, \ldots, \cos n\omega_0 t, \sin n\omega_0 t\}$ with a nonvanishing $\cos n\omega_0 t$ or $\sin n\omega_0 t$ term. This is discussed in Section V.1.

IV.5 JUMP RESONANCE

In Sec. IV.3, we pointed out that the contraction condition disallowed jump resonance. We discuss jump resonance in more detail in this section particularly because it bears on the question of how close to necessary are the conditions of Sec. IV.3. First, we should clarify what we mean here by jump resonance and nonuniqueness. The systems under consideration in the section have nonlinearities which are Lipschitzian. In Example II.1.2, we have seen that there are unique solutions to differential equations satisfying a Lipschitz condition. That is, from a given initial condition, one and only one (continuous) solution evolves in time. In the present section, uniqueness is relative to input functions,[†] not with respect to initial conditions. That is, a system satisfying conditions for uniqueness with respect to initial conditions can still have a non-unique periodic response to a sinusoidal input. The two different responses correspond to different initial conditions at any time. Saying that a system has a unique response to a given input usually means that, with two different initial conditions, the responses will asymptotically approach each other. When using a space of periodic functions, we do not explicitly focus attention on initial conditions or transient effects. This will be done in the chapter on stability.

Just as violation of the last condition in (IV.3-16) is a necessary condition for the existence of jump resonance for the exact system (IV.2-1), we obtain from (IV.4-13) a necessary condition for jump resonance for the approximate system (IV.4-11). In particular, for $P_n = P_1$, a necessary condition for nonunique solutions is

$$\max_{k=0,\pm1} \left| \frac{H_k}{1 + \frac{1}{2}(\alpha + \beta)H_k} \right| \tfrac{1}{2}(\beta + \alpha) \geq 1. \qquad \text{(IV.5-1)}$$

Suppose one does find nonunique solutions to (IV.4-11). The question immediately arises as to what that implies about the original system (IV.2-1). This question is clearly related to how close to necessary are the conditions of Sec. IV.3, a topic we only touched on in that section (we indicated what happens when there is a number H_k^{-1} on the real axis diameter of the circular disc but did not discuss the presence of numbers H_k^{-1} in the disc but not on the real axis diameter).

Fukuma and Matsubara[‡] have derived a criterion for the prediction of jump resonance with sinusoidal input by describing functions (equivalent linearization). For brevity of notation, we shall refer to this condition as d.f.c. (for describing function criterion). Recognizing that (IV.3-19) (hereafter in this section referred to as c.c. for contraction condition) is the key part of the sufficient condition for the absence of jump resonance, it is of interest to compare the d.f.c. and c.c. Two questions arise:

(i) Is it possible to satisfy the d.f.c. for jump resonance while also satisfy-

---

† In the context of differential equations, different input functions mean different right-hand sides of the differential equations.

‡ Also see Hatanaka.

ing the c.c. which disallows jump resonance? That is, does the c.c. ever show the d.f.c. to be incorrect?

(ii) If the c.c. is violated, is it always possible to find a nonlinearity with slopes between α and β such that jump resonance is predicted by the d.f.c.?

The object in (i) is to judge the d.f.c. by the c.c., and in (ii) to judge the c.c. by the d.f.c. The answer to both questions will be found to be "no." Of course, nothing will be proved mathematically regarding either the d.f.c. or the c.c. since the latter is only a sufficient condition and the former is heuristic. Furthermore, the class of inputs considered by the d.f.c. (sinusoids) is much less general than the class considered for the c.c. (periodic functions square integrable over a period). Nevertheless, the comparison is informative and suggestive.

The d.f.c. is concerned with the same feedback loop as described in Sec. IV.2. For simplicity, we shall restrict our attention to piecewise smooth nonlinearities with $\alpha = 0$, $\beta = 1$, i.e., where $n'(u)$ exists,

$$0 \leq n'(u) \leq 1. \tag{IV.5-2}$$

The linear operator L is defined by the transfer function $H(i\omega)$. There are some other assumptions made in Fukuma and Matsubara which we need not go into because the basic technique (describing functions) is itself heuristic and this section is intended to be suggestive, not definitive. The d.f.c. states that if the locus of $H^{-1}(i\omega)$ enters a particular circular disk on the complex plane, then jump resonance will occur for some E, the amplitude of a sinusoidal input to the nonlinearity in the describing function analysis. The circle is centered at $(c_1, 0)$ with radius ρ,

$$c_1 = -\frac{1}{\pi} \int_{-\pi/2}^{\pi/2} n'(E \sin \theta) \, d\theta \,, \tag{IV.5-3a}$$

$$\rho = \frac{1}{\pi} \left| \int_{-\pi/2}^{\pi/2} n'(E \sin \theta) \cos 2\theta \, d\theta \right|. \tag{IV.5-3b}$$

One considers the union over E of all the circular discs described by (IV.5-3) and sees whether the locus of $H^{-1}(i\omega)$ enters that region. If it does, jump resonance is predicted.

Although question (i) has actually already been answered by the discussion in connection with (IV.4-13) and (IV.5-1), it may be informative to also provide an answer by directly using the d.f.c. This is done simply by showing that the discs described by (IV.5-3) are always contained within the disc of Fig. IV.1 with $\alpha = 0$, $\beta = 1$. Then satisfaction of the c.c. precludes satisfaction of the c.c. From (IV.5-3),

$$-(c_1 - p) = \frac{1}{\pi} \int_{-\pi/2}^{\pi/2} n'(E \sin \theta)(1 + \cos 2\theta) \, d\theta$$

or

$$\frac{1}{\pi} \int_{-\pi/2}^{\pi/2} n'(E \sin \theta)(1 - \cos 2\theta) \, d\theta . \qquad (\text{IV.5-4})$$

In either case, use of (IV.5-2) implies that $-(c_1 - p) \leq 1$ or $(c_1 - p) \geq -1$. Since we also have $(c_1 + p) \leq 0$, the relationship between the circles is as shown in Fig. IV.2.

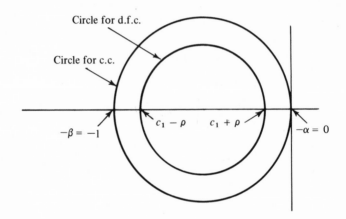

Figure IV.2 Relationship between circles

That the c.c. does not contradict the d.f.c. may add a little confidence to the describing function user. Of course, the c.c. is not known to be a necessary condition for the nonexistence of jump resonance and thus nothing has been mathematically proved regarding the d.f.c. It perhaps would have been more informative though more disheartening to the describing function user had we found out that the d.f.c. may be contradicted by the c.c. Along the lines of seeing where the describing function method fails, we shall see in Sec. VI.4 a case where the describing function method fails to indicate self-sustained oscillations.

Question (ii) may be restated as follows: If the locus of $H^{-1}(i\omega)$ enters the disc of radius $\frac{1}{2}$ with center at $(-1/2, 0)$, does it also enter a disc that predicts jump resonance for a nonlinearity satisfying (IV.5-2)? Note that the c.c. is valid for any nonlinearity satisfying (IV.5-2); so, for this question, one is free to shape the nonlinearities in looking for jump resonance.

It can be shown that the jump resonance regions for saturation and deadzone[†] include the jump regions for all other nonlinearities satisfying (IV.5-2). This is determined by maximizing the radius ρ for different c_i's [see (IV.5-3)].[‡] In Fig. IV.3, the saturation jump region is shown to be inside the disc of the c.c. Figure IV.3 indicates that one may violate the c.c. by having the locus of $H^{-1}(i\omega)$ enter the circle without also satisfying the d.f.c. for jump resonance. This should be carefully interpreted. First of all, as repeatedly mentioned, the d.f.c. is not rigorous. However, although the d.f.c. is generally approximate, it can be exact in predicting jump resonance at frequency ω_0 if $G(ik\omega_0) = 0$ for $k > 1$ so that a large portion of the c.c. circle may be filled with jump resonance points. Second, it is only for sinusoidal inputs. There may be other periodic inputs causing jump phenomena (though it has not been shown how to sharpen the c.c. if only sinusoidal inputs are considered).

Considering the above discussion, this section should serve the purpose of directing the reader's attention to some interesting open questions.

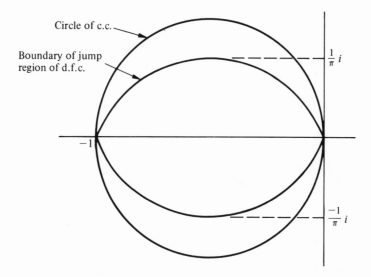

Figure IV.3 Comparison of jump regions

IV.6 IMPROVING ACCURACY

In the last section, we speculated on the consequences of nonsatisfaction of condition (IV.3-18). Now suppose that (IV.3-18) is satisfied, but one is dis-

† Saturation and deadzone are defined in a footnote in Sec. III.4.
‡ See the Appendix to this chapter.

satisfied with the bound on $\|x^* - x_0\|$ [(IV.3-4) with $m = 0$] with x_0 an equivalent linearization approximation. Condition (IV.3-4) itself immediately suggests how to improve the accuracy: Iterate with (IV.3-3). The bound could theoretically be made as small as desired. However, after one or two iterations, the procedure may become quite complicated (which may not be a drawback if a computer is used). Another way of improving accuracy is to use a more sophisticated approximation, namely (IV.4-11) with $n > 1$. If (IV.4-11) is used then, with $c = \frac{1}{2}(\alpha + \beta)$, with x_0 now depending on P_n, and recognizing that the order of P_n and L can be interchanged,

$$\begin{aligned}
\|x^* - x_0\| &\leq \frac{\|(I + cL)^{-1}[-LNx_0 + P_nLNx_0]\|}{1 - \gamma} \\
&= \frac{\|(I + cL)^{-1}L(I - P_n)Nx_0\|}{1 - \gamma} \\
&\leq \frac{1}{1 - \gamma} \sup_{|k| > n} \left|\frac{H_k}{1 + cH_k}\right| \cdot \|Nx_0\|.
\end{aligned} \tag{IV.6-1}$$

If $n(0) = 0$ and $H_k \to 0$ as $k \to \infty$ then the last bound also goes to zero as $n \to \infty$ since it may be verified that for all n

$$\|Nx_0\| \leq \max\{\beta, |\alpha|\} \frac{1}{1 - \gamma} \sup_k \left|\frac{1}{1 + cH_k}\right| \cdot \|r\|. \tag{IV.6-2}$$

Exercise IV.6.1

Derive the equivalent of (IV.6-2) for $n(0) \neq 0$.

IV.7 NOTES

Information on Fourier series may be found in Kaplan and Tolstov. Our treatment of Hilbert spaces follows Kantorovich and Akilov; the reader may also find Porter useful here. The contraction mapping analysis of the feedback system is based on Sandberg [3] and [4]. One motivation for equivalent linearization has been a minimization of the type done in Sec. IV.4 without relation to a contraction mapping analysis (e.g., Hsu and Meyer, Sec. 6.4). It then becomes very general but theoretical justification is known to be available when contraction conditions are satisfied (otherwise it remains a useful heuristic tool). A discussion of equations of the form of (IV.4-11) with $n = 2$ is given in Hsu and Meyer in their chapter on dual-input describing functions. For computational considerations in connection with (IV.4-11) in the form of a differential equation and large n, see Urabe and Reiter. The book by Gelb and Vander-

Velde is devoted almost entirely to the application of the describing function method.

Related contraction mapping analyses of nonlinear feedback systems for transient analysis and random inputs are given in Bickart and Holtzman [3], respectively. Kyong adapted the approach of Sec. IV.5 and the Appendix to the study of nonunique responses to random inputs. Both Holtzman [3] and Kyong are concerned with statistical linearization, which is the random function counterpart to equivalent linearization for periodic functions.

APPENDIX

Boundary of Jump Resonance Region For Section IV.5

Referring to (IV.5-3) we wish to maximize ρ for c_1's between 0 and -1 [in view of (IV.5-2)]. Let

$$g(\theta) = \tfrac{1}{2}[n'(E \sin \theta) + n'(E \sin (-\theta))]. \qquad \text{(IV.A-1)}$$

Then consider choosing $g(\theta)$ to maximize

$$\rho = \frac{2}{\pi}\left| \int_0^{\pi/2} g(\theta) \cos 2\theta \, d\theta \right| \qquad \text{(IV.A-2)}$$

subject to

$$0 \le g(\theta) \le 1, \qquad \theta \in [0, \pi/2], \qquad \text{(IV.A-3)}$$

$$\int_0^{\pi/2} g(\theta) \, d\theta = \tfrac{1}{2}\lambda\pi \qquad \text{(IV.A-4)}$$

for $\lambda \in [0, 1]$. In Fukuma and Matsubara, it is recognized by inspection that the maximizing $g(\theta)$'s are those represented by saturation and deadzone. That is,

$$g(\theta) = 1 \qquad \theta \in [0, \lambda\pi/2]$$
$$\quad\;\; = 0 \qquad \theta \in (\lambda\pi/2, \pi/2] \tag{IV.A-5}$$

or

$$g(\theta) = 0 \qquad \theta \in [0, (1 - \lambda)\pi/2)$$
$$\quad\;\; = 1 \qquad \theta \in [(1 - \lambda)\pi/2, \pi/2]. \tag{IV.A-6}$$

Those familiar with Pontryagin's maximum principle (see, e.g., Hsu and Meyer, Chapter 14) may be interested in seeing it applied to the above optimization problem. Let

$$\dot{x}_1 = g(\theta), \qquad\quad x_1(0) = 0,$$
$$\dot{x}_2 = g(\theta)\cos 2\theta, \quad x_2(0) = 0. \tag{IV.A-7}$$

Consider maximizing $x_2(\pi/2)$ subject to the constraint on the control

$$0 \le g(\theta) \le 1, \qquad \theta \in [0, \pi/2] \tag{IV.A-8}$$

and the fixed endpoint

$$x_1(\pi/2) = \lambda\pi/2. \tag{IV.A-9}$$

The Hamiltonian is

$$H = p_1 g(\theta) + p_2 g(\theta)\cos 2\theta \tag{IV.A-10}$$

with $\dot{p}_1 = \dot{p}_2 = 0$ and $p_2 \ge 0$. We must have $p_2 > 0$ in order that (IV.A-9) be satisfied for $\lambda \in (0, 1)$. Letting $p_2 = 1$ and maximizing H with respect to $g(\theta)$ subject to (IV.A-8) yields

$$g(\theta) = \tfrac{1}{2} + \tfrac{1}{2}\operatorname{sgn}[p_1 + \cos 2\theta]. \tag{IV.A-11}$$

The number p_1 is chosen to satisfy (IV.A-9). The $g(\theta)$ of (IV.A-11) corresponds to saturation. Deadzone would be obtained if we minimized $x_2(\pi/2)$ instead of maximizing it (recall we actually wish to maximize $|x_2(\pi/2)|$).

The maximum principle is only a necessary condition that an optimal control must satisfy (if an optimal control exists). However, it may be shown that the optimum does exist for this problem and that the control obtained above is indeed optimal.

V

Periodic Solutions–Local

Contraction

The contraction result of the last chapter is very elegant with a most simple geometric interpretation. The price paid for the elegance is that it is rather restrictive. The restrictiveness is largely a consequence of requiring the mapping to be a global contraction, a strong condition. This chapter relaxes the global contraction requirement by using Theorem II.3.2 and then by using a new argument when the CMT fails.

Instead of continuing to concentrate on the general feedback configuration of the last two chapters, we shall use some special techniques that are useful with differential equations. One becomes better prepared to attack nonlinear problems as he fills his bag of tools; facility with both transform methods and differential equations is most helpful. We shall, in fact, use both in this chapter.

V.1 THE CHOICE OF BANACH SPACE

It should be clear from previous chapters that spaces of square integrable functions $(L_2(-\infty, \infty), B_T)$ are most convenient to work with largely because they are Hilbert spaces and because conditions can be stated in the frequency domain. In this chapter we shall primarily use C_T, the Banach space of real-valued continuous n-vector T-periodic functions with norm

$$\|x\| = \max_{i=1,\ldots,n} \max_t |x_i(t)| \qquad (V.1\text{-}1)$$

(subscript denotes component of vector).

In using this space, we lose some of the simplicity associated with Hilbert spaces. As compensation, we have the following advantages. First of all, the nondecreasing functions needed for Theorem II.3.2 are found quite easily; $\|N'(x)\|$ is usually a more "natural" function of $\|x\|$ in the C_T norm than in the B_T norm. Secondly, the C_T norm is often more desirable in error analysis (the error between an approximation and an exact solution) than is the B_T norm. The C_T norm provides a bound on the magnitude of the error while the B_T norm gives the rms error. Finally, the conditions imposed on a nonlinear function to map B_T into itself are quite restrictive. The nonlinear operator of interest $y = Nx$ is often defined by the real function of real variables

$$y(t) = n(x(t), t). \qquad (V.1\text{-}2)$$

Under some mild conditions, it may be shown that a necessary condition for this operator to map $L_2(0, T)$ into itself is that for some $b > 0$ and some $a \in L_2(0, T)$

$$|n(t, u)| \leq a(t) + b|u|, \qquad t \in [0, T] \qquad (V.1\text{-}3)$$

(Krasnosel'skii, p. 27). The requirement that a continuous function be mapped into a continuous function is often more convenient. For example, polynomials do not satisfy (V.1-3). However, in some cases one may focus attention on some subset and less restrictive requirements on the nonlinearity might be imposed. Also, the nonlinearities for many control engineering problems are Lipschitzian (as in the last chapter), and B_T can be used.

The above should make it clear that we are not drawing any hard and fast lines regarding where to use which space. Rather our intent is to mention some considerations that may be taken into account and to illustrate use of different spaces. Having committed ourselves in this chapter to using C_T, we wish to discuss some implications of using that space.

We closed Sec. IV.4 by pointing out that in C_T the best trigonometric polynomial approximation is not generally the partial sum of the Fourier series as it is in B_T. Let us discuss partial sums in C_T. Using the Fourier series representation for $x \in C_T$ (in the rest of this section, assume for simplicity that x is a scalar function)

$$x(t) \sim \frac{a_0}{2} + \sum_{k=1}^{\infty} (a_k \cos k\omega t + b_k \sin k\omega t), \qquad \omega = \frac{T}{2\pi}, \qquad (V.1\text{-}4)$$

the operator P_n given by

$$(P_n x)(t) = \frac{a_0}{2} + \sum_{k=1}^{n} (a_k \cos k\omega t + b_k \sin k\omega t) \qquad \text{(V.1-5)}$$

is a projection also in C_T. This may be seen by referring to the definition of projections for linear spaces in Sec. IV.1, and decomposing C_T into the subspace spanned by $\{1, \cos \omega t, \sin \omega t, \ldots, \cos n\omega t, \sin n\omega t\}$ (the space of trigonometric polynomials of degree $\leq n$) and the subspace of continuous periodic functions with vanishing first $(2n + 1)$ Fourier coefficients $a_0, \ldots, a_n, b_1, \ldots, b_n$.

The projection operator P_n in C_T does not satisfy $\|P_n\| \leq 1$ as projections in a Hilbert space must [see Exercise IV.1.2(e)]. This can be seen easily by example. Consider the function $x(t) = |\sin t|$ which has the following Fourier series (which may be shown to be absolutely and uniformly convergent; Tolstov, p. 26):

$$x(t) = \frac{2}{\pi} - \frac{4}{\pi}\left(\frac{\cos 2t}{3} + \frac{\cos 4t}{15} + \frac{\cos 6t}{35} + \cdots\right). \qquad \text{(V.1-6)}$$

Then

$$(P_2 x)(t) = \frac{2}{\pi} - \frac{4}{\pi}\frac{\cos 2t}{3} \qquad \text{(V.1-7)}$$

which leads to the following situation:

$$\|x\| = 1, \qquad \|P_2 x\| = \frac{10}{3\pi} \Rightarrow \|P_2\| \geq \frac{10}{3\pi} > 1. \qquad \text{(V.1-8)}$$

In fact, it may be shown that in general

$$\|P_n\| > \frac{4 \ln n}{\pi^2} \qquad \text{(V.1-9)}$$

(Cheney, p. 212) so that $P_n x$ could be a very poor approximation to x if maximum magnitudes are considered (it is possible that $\|x - P_n x\| \to \infty$ as $n \to \infty$). Incidentally, this is closely related to the existence of divergent Fourier series mentioned in Sec. IV.1.

While $P_n x$ could be a poor approximation to x in C_T, it is not necessarily so. In particular, we are interested in periodic solutions to differential equations which have continuous derivatives which imply that they are continuous and piecewise smooth. Such periodic functions have uniformly convergent Fourier series (see, e.g., Tolstov, p. 81) which implies $\lim_{n\to\infty} \|x - P_n x\| = 0$ (compare with Theorem IV.1.1; we have added here the condition of continuity). Furthermore, in our applications (e.g., in Sec. V.3) we need not be concerned with all

$\|x - P_n x\|$ but only with $\|x_0 - P_n x_0\|$ for given n and some special x_0's and these specific quantities will be bounded. Further discussion of trigonometric polynomials is given in the appendix to this chapter.

Having discussed some generalities about the periodic functions in C_T we now wish to discuss a class of systems which may exhibit solutions in C_T.

V.2 THE SYSTEM UNDER CONSIDERATION

The system to be studied is described by the n-vector differential equation

$$\frac{dx(t)}{dt} = A(t)x(t) + n(x(t), t) \tag{V.2-1}$$

with $A(t)$ and $n(u, t)$ both T-periodic in t and $n(x(t), t)$ continuous in t if $x(t)$ is. All vectors are real. It is convenient to convert this differential equation into an equivalent integral equation. First we recall some properties of the linear differential equation

$$\frac{dx(t)}{dt} = A(t)x(t) + f(t) \tag{V.2-2}$$

with $A(t)$ and $f(t)$ continuous. Existence and uniqueness of solutions to (V.2-2) are easily settled by arguments given in Sec. II.1. The solution to (V.2-2) is given by

$$x(t) = Y(t)x(0) + \int_0^t Y(t)Y^{-1}(s)f(s)\,ds \tag{V.2-3}$$

where $Y(t)$ is a fundamental solution matrix $[Y(t)Y^{-1}(s) = \Phi(t, s)$, the transition matrix]. Some familiar properties of $Y(t)$ are listed below for reference:

$$\frac{d}{dt} Y(t) = A(t)Y(t), \quad Y(0) = I,$$

$$\det Y(t) \neq 0. \tag{V.2-4}$$

Letting $A(t)$ and $f(t)$ in (V.2-2) both be periodic of period T, we have the following theorem.

Theorem V.2.1

A necessary and sufficient condition in order that, for any f of period T, the differential equation (V.2-2) has a T-periodic solution is that the corresponding homogeneous system [(V.2-2) with $f = 0$] has no nontrivial T-periodic solutions (nontrivial = nonzero).

Proof. First we show that a solution is T-periodic if and only if $x(T) = x(0)$. A solution with period T obviously satisfies this condition. Conversely, if the condition is satisfied, the solutions $x(t + T)$ and $x(t)$ coincide for $t = 0$. The uniqueness of the solutions imply they coincide for any t and thus $x(t)$ has period T. Now, from (V.2-3) with $x(T) = x(0)$,

$$[I - Y(T)]x(0) = \int_0^T Y(T)Y^{-1}(s)f(s)\,ds. \tag{V.2-5}$$

The condition that (V.2-5) has a solution for $x(0)$ for any f is that

$$\det[I - Y(T)] \neq 0. \tag{V.2-6}$$

This means that the equation

$$Y(T)x(0) = x(0) \tag{V.2-7}$$

has no solution other than $x(0) = 0$ which in turn disallows nontrival solutions of period T to the homogeneous system.

Remark. From the preceding proof, it is seen that the existence of $[I - Y(T)]^{-1}$ is equivalent to the nonexistence of nontrival T-periodic solutions to $\dot{x}(t) = A(t)x(t)$.

The next theorem shows how to convert (V.2-2) into an integral equation convenient for examination of periodic solutions.

Theorem V.2.2

If $[I - Y(T)]^{-1}$ exists, the unique T-periodic solutions of (V.2-2) can be expressed as

$$x(t) = \int_0^T G(t, s)f(s)\,ds, \tag{V.2-8}$$

where

$$G(t, s) = \begin{cases} Y(t)[I - Y(T)]^{-1}Y^{-1}(s) & 0 \leq s \leq t \leq T \\ Y(t + T)[I - Y(T)]^{-1}Y^{-1}(s) & 0 \leq t < s \leq T. \end{cases} \tag{V.2-9}$$

Proof. From $x(T) = x(0)$, we have

$$[I - Y(T)]x(0) = \int_0^T Y(T)Y^{-1}(s)f(s)\,ds. \tag{V.2-10}$$

Using the invertibility of $[I - Y(T)]$ to solve for $x(0)$ in (V.2-3),

$$x(t) = Y(t)[I - Y(T)]^{-1} \int_0^T Y(T) Y^{-1}(s) f(s) \, ds$$

$$+ \int_0^t Y(t) Y^{-1}(s) f(s) \, ds . \tag{V.2-11}$$

The remainder of the proof is left as

Exercise V.2.1

Obtain (V.2-8) and (V.2-9) from (V.2-11).

The assumption of the nonsingularity of the matrix $[I - Y(T)]$ is sometimes referred to as being the *noncritical case*. The more difficult critical case is discussed in the next chapter.

Returning to (V.2-1), we form the nonlinear integral equation

$$x(t) = \int_0^T G(t, s) n(x(s), s) \, ds. \tag{V.2-12}$$

Any continuous solution of (V.2-12) represents a periodic solution to (V.2-1), since we obtain by differentiating (V.2-12), referring to (V.2-11), and using (V.2-4),

$$\frac{dx(t)}{dt} = A(t) Y(t)[I - Y(T)]^{-1} \int_0^T Y(T) Y^{-1}(s) n(x(s), s) \, ds$$

$$+ Y(t) Y^{-1}(t) n(x(t), t)$$

$$+ \int_0^t A(t) Y(t) Y^{-1}(s) n(x(s), s) \, ds$$

$$= A(t) x(t) + n(x(t), t) . \tag{V.2-13}$$

The problem of finding periodic solutions of (V.2-1) is thus reduced to finding a continuous solution to a nonlinear integral equation. We need only consider the interval $[0, T]$ because $G(t, s)$ was constructed so that $x(T) = x(0)$. That is, we can actually work in the space of continuous functions on one period rather than the space of continuous periodic functions. We shall next use Theorem II.3.2 to find continuous solutions of (V.2-12). We consider the righthand side of (V.2-12) to define an operator LN with L the linear integral operator. We then look for fixed points of LN. The function x_0 can be an approximate solution of (V.2-12). With x_0 obtained by harmonic balance,

$$x_0 = \tilde{L} N(x_0), \tag{V.2-14}$$

where, if

$$(Lx)(t) \sim \frac{a_0}{2} + \sum_{k=1}^{\infty} (a_k \cos k\omega t + b_k \sin k\omega t), \tag{V.2-15}$$

then

$$(\tilde{L}x)(t) = a_1 \cos \omega t + b_1 \sin \omega t \tag{V.2-16}$$

(the a_k, b_k are vectors, in general). Hence, if $a_0 = 0$, $\tilde{L}x = P_1 Lx$. Note that, if we consider $x = LN(x)$ to represent a feedback system, x represents the output (rather than the error as in Chapters III and IV) and the forcing function is absorbed in $N(x)$ (rather than displayed separately as in Chapters III and IV).

V.3 EXAMPLE: DUFFING'S EQUATION

An example will best explain the application of Theorem II.3.2. Consider Duffing's equation

$$\ddot{y}(t) + ay(t) + b[y(t)]^3 = f \cos \omega t \qquad (a > 0) \tag{V.3-1}$$

already discussed in the Introduction. It has the feedback equivalent shown in Figure V.1 [in which we subscripted the input frequency to distinguish it from the argument of the transfer function $H(i\omega)$].

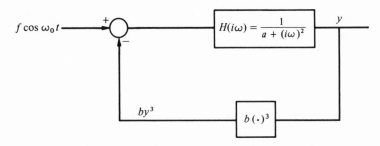

Figure V.1 Feedback equivalent of Duffing's equation

It is clear that this feedback system does not satisfy the conditions of the last chapter because the cubic does not satisfy a (global) Lipschitz condition. In the Introduction, we saw that harmonic balance indicates that

$$y(t) = A \cos \omega t \tag{V.3-2}$$

with

$$\tfrac{3}{4}bA^3 + (a - \omega^2)A - f = 0 \tag{V.3-3}$$

is an approximate solution of the equation. We wish to verify if there is an exact solution near this approximation, and if so, how near.

Letting

$$x = \begin{pmatrix} x_1 \\ x_2 \end{pmatrix} \tag{V.3-4}$$

$$x_1 = y \tag{V.3-5}$$

$$x_2 = \dot{x}_1 \tag{V.3-6}$$

the corresponding vector differential equation is

$$\dot{x}(t) = \begin{pmatrix} 0 & 1 \\ -a & 0 \end{pmatrix} x(t) + \begin{pmatrix} 0 \\ f \cos \omega t - b x_1^3(t) \end{pmatrix}. \tag{V.3-7}$$

The fundamental solution matrix for

$$\dot{x}(t) = \begin{pmatrix} 0 & 1 \\ -a & 0 \end{pmatrix} x(t) \tag{V.3-8}$$

is

$$Y(t) = \begin{pmatrix} \cos \sqrt{a}\, t & \dfrac{\sin \sqrt{a}\, t}{\sqrt{a}} \\ -\sqrt{a}\, \sin \sqrt{a}\, t & \cos \sqrt{a}\, t \end{pmatrix}. \tag{V.3-9}$$

$G(t, s)$ is given by

$G(t, s) =$

$$\begin{cases} \dfrac{1}{2 \sin \left(\frac{\sqrt{a}T}{2} \right)} \left[\begin{array}{cc} \sin \sqrt{a} \left(\frac{T}{2} - t + s \right) & \frac{1}{\sqrt{a}} \cos \sqrt{a} \left(\frac{T}{2} - t + s \right) \\ -\sqrt{a} \cos \sqrt{a} \left(\frac{T}{2} - t + s \right) & \sin \sqrt{a} \left(\frac{T}{2} - t + s \right) \end{array} \right] \\ \qquad\qquad\qquad\qquad\qquad\qquad\qquad\qquad 0 \le s \le t \le T \\[2mm] \dfrac{1}{2 \sin \left(\frac{\sqrt{a}T}{2} \right)} \left[\begin{array}{cc} -\sin \sqrt{a} \left(\frac{T}{2} + t - s \right) & \frac{1}{\sqrt{a}} \cos \sqrt{a} \left(\frac{T}{2} + t - s \right) \\ -\sqrt{a} \cos \sqrt{a} \left(\frac{T}{2} + t - s \right) & -\sin \sqrt{a} \left(\frac{T}{2} + t - s \right) \end{array} \right] \\ \qquad\qquad\qquad\qquad\qquad\qquad\qquad\qquad 0 \le t < s \le T. \end{cases} \tag{V.3-10}$$

The integral operator of interest, $y = LN(x)$, is defined by

$$y(t) = \int_0^T G(t, s) \begin{pmatrix} 0 \\ f \cos \omega s - b x_1^3(s) \end{pmatrix} ds. \tag{V.3-11}$$

Note that in the above, y is just used to define $LN(x)$; it is not the solution of (V.3-1). The approximate solution (obtained from harmonic balance) is x_0, i.e.,

$$x_0(t) = \begin{pmatrix} x_{01}(t) \\ x_{02}(t) \end{pmatrix} = \begin{pmatrix} A \cos \omega t \\ -\omega A \sin \omega t \end{pmatrix} \tag{V.3-12}$$

with ω and A being related by (V.3-3).

With the norm of (V.1-1), it is easily verified that[†]

$$\|LN(x_0) - x_0\| \le T \max_{i=1,2} \max_{t,s \in [0,T]} |G_{i2}(t, s)| \frac{|bA^3|}{4}$$

$$\le Tc \frac{|bA^3|}{4}, \tag{V.3-13}$$

where

$$c = \frac{1}{2 |\sin (\sqrt{a}\, T/2)|} \max \{1, 1/\sqrt{a}\}. \tag{V.3-14}$$

The derivative operator $z = (LN)'(x)\bar{x} = LN'(x)\bar{x}$ is given by

$$z(t) = \int_0^T G(t, s) \begin{pmatrix} 0 \\ -3b x_1^2(s)\bar{x}_1(s) \end{pmatrix} ds$$

$$= \int_0^T \begin{bmatrix} -G_{12}(t, s)3b x_1^2(s) & 0 \\ -G_{22}(t, s)3b x_1^2(s) & 0 \end{bmatrix} \begin{bmatrix} \bar{x}_1(s) \\ \bar{x}_2(s) \end{bmatrix} ds. \tag{V.3-15}$$

Exercise V.3.1

Verify (V.3-13) and (V.3-15).

The norm of the derivative operator is bounded as follows:

[†] With zero mean values and with A a constant matrix,

$$P_1 \dot{x}_0(t) = P_1 A x_0(t) + P_1 n(x_0(t), t)$$
$$= A P_1 x_0(t) + P_1 n(x_0(t), t),$$
$$x_0(t) = \int_0^T G(t, s) P_1 n(x_0(s), s)\, ds.$$

Hence,

$$(LNx_0 - x_0)(t) = \int_0^T G(t, s) \big[n(x_0(s), s) - P_1 n(x_0(s), s) \big]\, ds.$$

$$
\begin{aligned}
\|LN'(x)\| &\leq \max_{i=1,2} \max_{t \in [0,T]} \int_0^T |G_{i2}(t, s) 3bx_1^2(s)| \, ds \\
&\leq 3|b|cT \left(\max_{t \in [0,T]} |x_1(t)| \right)^2 \\
&\leq 3|b|cT \left(\max_{t \in [0,T]} |x_{01}(t)| + \max_{t \in [0,T]} |x_1(t) - x_{01}(t)| \right)^2 \\
&\leq 3|b|cT(|A| + \|x - x_0\|)^2 \, . \qquad\qquad\qquad\qquad \text{(V.3-16)}
\end{aligned}
$$

We may note that the above bounds on norms may be slightly strengthened by closer analysis (e.g., by using)

$$
\max_{t \in [0,T]} \int_0^T |G_{i2}(t, s)| \, ds
$$

instead of

$$
T \max_{t, s \in [0,T]} |G_{i2}(t, s)|.
$$

One who is interested in the sharpest results would do this. For illustrative purposes here, we shall be satisfied with the simpler bounds.

The right-hand side of (V.3-16) defines the nondecreasing function g such that

$$
\|LN'(x)\| \leq g(\|x - x_0\|). \qquad\qquad\qquad\qquad \text{(V.3-17)}
$$

To use Theorem II.3.2, let

$$
\Omega = \left\{ x : \|x - x_0\| \leq \frac{k}{1 - \gamma} \right\}, \qquad\qquad\qquad\qquad \text{(V.3-18)}
$$

where

$$
k = cT \frac{|bA^3|}{4} \qquad\qquad\qquad\qquad \text{(V.3-19)}
$$

[see (V.3-13)]. If a $\gamma \in [0, 1)$ can be found satisfying

$$
3|b|cT \left(|A| + \frac{cT \, (|bA^3|/4)}{1 - \gamma} \right)^2 \leq \gamma, \qquad\qquad\qquad\qquad \text{(V.3-20)}
$$

then there is an $x^* \in \Omega$ such that $x^* = LN(x^*)$, i.e., Duffing's equation has a periodic solution in the neighborhood of the approximation obtained by harmonic balance.

Consider $a, f,$ and ω fixed and $f \neq 0$. Since, for any $\gamma \in [0, 1)$

$$\lim_{b \to 0} 3|b|cT\left(|A| + \frac{cT\left(|bA^3|/4\right)}{1-\gamma}\right)^2 = 0 \tag{V.3-21}$$

it is seen that for b sufficiently small, there will be a $\gamma \in [0, 1)$ satisfying (V. 3-20) (and thus a periodic solution neighboring the approximation).

To verify (V.3-21), recall that A depends on b, so let

$$
\begin{aligned}
z &= 3|b|cT\left[|A| + \frac{cT(|bA^3|/4)}{1-\gamma}\right]^2 \\
&= 3cT\left\{|bA^2| + \frac{(cT/2)|bA^2|^2}{1-\gamma} + \left(\frac{cT}{1-\gamma}\right)^2 \frac{|bA^2|^3}{16}\right\}.
\end{aligned}
\tag{V.3-22}
$$

Then (V.3-21) follows from (V.3-3) which implies

$$\lim_{b \to 0} |bA^2| = 0 \tag{V.3-23}$$

(recognizing that with $b = 0$, A is the amplitude of an oscillation of a linear differential equation). Note that while the result has been stated as an asymptotic result, it is possible to determine quantitatively what is meant by "sufficiently small." We shall actually give some quantitative results in later sections of this chapter.

V.4 SPECIAL CASES

In many cases it is not necessary to convert the original differential equation into a vector integral equation. For example, let the system be described by the scalar differential equation

$$a_0 x^{(m)}(t) + a_1 x^{(m-1)}(t) + \cdots + a_m x(t) = n(x(t), t), \tag{V.4-1}$$

where the $a_0, \ldots, a_m$ are real constants, $a_0 \neq 0$, and the real-valued function $n(x(t), t)$ is T-periodic in t and is continuous in t when $x(t)$ is. A periodic solution to (V.4-1) will, under broad conditions, satisfy the following integral equation

$$x(t) = \int_0^T W_T(t - u)n(x(u), u) \, du \tag{V.4-2}$$

where $W_T(t)$ is an inverse finite Fourier transform. We will not go into details concerning finite Fourier transforms; rather we refer the reader to a comprehensive discussion in Chapter 4 of Kaplan. We will, however, illustrate in the following example how to obtain $W_T(t)$.

Duffing's equation (V.3-1), is clearly of the form (V.4-1) and the integral equation corresponding to (V.4-2) is

$$x(t) = \frac{1}{2\sqrt{a}\,\sin\left(\sqrt{a}\,T/2\right)} \int_0^T \cos\sqrt{a}\left(\frac{T}{2} - t + s\right)\left[f\cos\omega s - bx^3(s)\right] ds$$

$$0 \le s \le t \le T,$$

$$= \frac{1}{2\sqrt{a}\,\sin\left(\sqrt{a}\,T/2\right)} \int_0^T \cos\sqrt{a}\left(\frac{T}{2} + t - s\right)\left[f\cos\omega s - bx^3(s)\right] ds$$

$$0 \le t < s \le T.$$
$$\text{(V.4-3)}$$

where now $x(t) = y(t)$, the solution to (V.3-1), not the vector with components $y(t)$, $\dot{y}(t)$ as in (V.3-5) and (V.3-6). To derive (V.4-3), we first find the finite Fourier transform $Y(ik\omega)$ associated with the differential equation (V.4-1),

$$Y(ik\omega) = \frac{1}{a_0(ik\omega)^m + \cdots + a_m} \tag{V.4-4}$$

which, for Duffing's equation, reduces to $Y(ik\omega) = \left[(ik\omega)^2 + a\right]^{-1}$. Then the inverse finite Fourier transform of $Y(ik\omega)$ is determined for $0 \le t < T$ (e.g., by using entry 11 on the table on p. 210 of Kaplan). The inverse finite Fourier transform is the (periodic) function $W_T(t)$ needed for (V.4-3). The equation (V.4-3) is consistent with and also can be derived from (V.3-11). However, the point of this section is to avoid matrix manipulations such as evaluating $G(t, s)$.

We can reapply Theorem II.3.2 to the integral operation defined by the right side of (V.4-3) (operating on the space of real-valued *scalar* continuous functions on $[0, T]$). We omit the details (or leave as exercise for the reader) because they exactly parallel those of Sec. V.3 except they are simpler in not requiring dealing with matrices. We get the same condition (V.3-20) except that now

$$c = \frac{1}{2|\sin\left(\sqrt{a}\,T/2\right)|}\frac{1}{\sqrt{a}}. \tag{V.4-5}$$

Using (V.4-5) instead of (V.3-14) represents a relaxation of condition (V.3-20) and may yield smaller $\gamma's$ since the c of (V.4-5) may be smaller than the c of (V.3-14). We thus can get a smaller bound on $\|x^* - x_0\|$ but one should keep in mind that

$$\|x^* - x_0\| = \max_{t \in [0,T]} \max\{|y^*(t) - y_0(t)|, |\dot{y}^*(t) - \dot{y}_0(t)|\}$$

$$\text{(Section V.3)}$$

$$\|x^* - x_0\| = \max_{t \in [0,T]} |y^*(t) - y_0(t)| \qquad \text{(Section V.4)}$$
$$\text{(V.4-6)}$$

where y^* and y_0 are the exact and approximate solutions of (V.3-1), respectively. Though the first norm in (V.4-6) could give more information, it is usually of interest to obtain the smallest bound on the second norm. Furthermore, the range of applicability of the contraction condition (V.3-20) is extended with a smaller c. Hence, working with the scalar integral equation can yield more valuable results as well as being simpler. As the order of the differential equation (V.4-1) increases it clearly becomes more advantageous to avoid the use of matrices.

Bergen and Franks study the application of Theorem II.3.2 to feedback systems with a convolution representation for the linear operator related to that of Chapters III and IV.

V.5 JUMP RESONANCE

In this section, we use condition (V.3-20) with the Sec. V.4 value of c given by (V.4-5). Before doing this, let us consider (V.3-3) which we rewrite here for convenience

$$\tfrac{3}{4}bA^3 + (a - \omega^2)A - f = 0. \tag{V.5-1}$$

(V.5-1) represents a condition relating the amplitude and frequency of an approximate solution of Duffing's equation, (V.3-1). It is of interest to plot a frequency response corresponding to (V.5-1), i.e., a plot of $|A|$ versus ω. To do this, it is more convenient to solve for ω as a function of A because it turns out that for certain values of ω there is more that one real value of A (i.e., there is jump resonance indicated). Solving (V.5-1) for ω yields

$$\omega^2 = a + \tfrac{3}{4}bA^2 - \frac{f}{A}. \tag{V.5-2}$$

Figure V.2 shows how the frequency response plot may be constructed for $f = 2, a = 10, b = 0.05$. Referring to the discussion of Duffing's equation in the Introduction, $b > 0$ corresponds to a hard spring. With a soft spring ($b < 0$), the frequency response curves would be bent to the left. From Figure V.2c, it is seen that, for ω larger than a number exceeding $\sqrt{a}$, there are three amplitudes $|A|$ corresponding to each ω; this is the jump reasonance predicted by harmonic balance. That jump reasonance does actually occur has been experimentally verified in some related cases (see, e.g., McLachlan, p. 60).

Figure V.3 show the frequency response plot of A versus ω; also indicated are those portions of the frequency response curve for which the contraction condition (V.3-20) with the c of (V.4-5) is satisfied. The contraction constants γ and errors e are indicated at some points. By error we mean

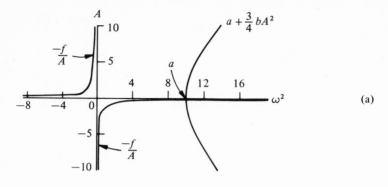

(a)

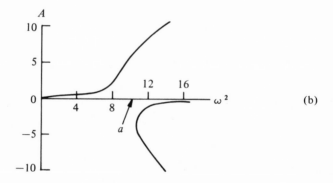

(b)

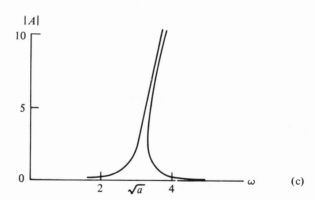

(c)

Figure V.2 Construction of frequency response plot ($f = 2$, $a = 10$, $b = 0.05$)

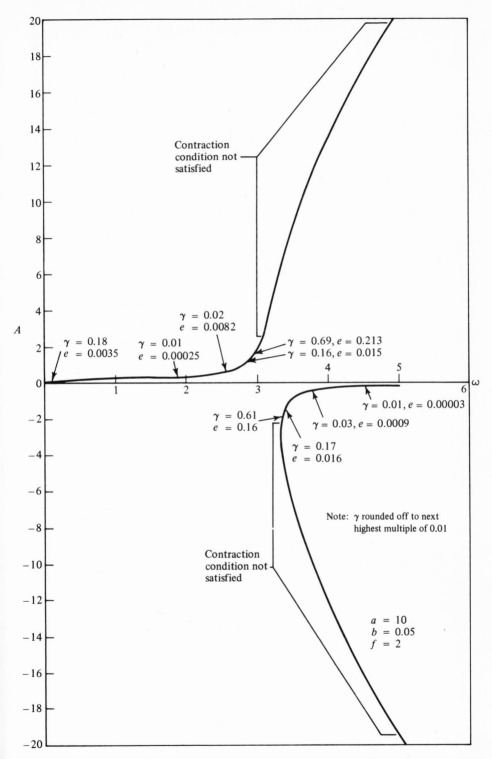

Figure V.3 Satisfaction of contraction condition on frequency response plot

$$e \geq \max_t |y^*(t) - y_0(t)|, \tag{V.5-3}$$

where y^* and y_0 are the exact and (harmonic balance) approximate periodic solutions, respectively. Figure V.3 shows that the contraction mapping analysis does not verify the existence of jump resonance for this case. This, of course, does not mean that the harmonic balance prediction of jump resonance is false since the CMT gives only sufficient conditions. The reason for the contraction mapping analysis not being applicable to distinct fixed points of the same frequency† (if there are indeed such fixed points associated with each harmonic balance approximation) is based on Theorem II.3.3.

V.6 EXISTENCE OF SUBHARMONICS

In the Introduction, we mentioned that the method of harmonic balance may predict subharmonic solutions to Duffing's equation. If we put

$$y(t) = A_1 \cos \omega t + A_3 \cos 3\omega t \tag{V.6-1}$$

into

$$\ddot{y}(t) + ay(t) + b[y(t)]^3 = f \cos 3\omega t \qquad (a, b, f > 0) \tag{V.6-2}$$

and balance terms with frequencies ω and 3ω, we find that the following equations must be satisfied:

$$\begin{aligned}
(a - \omega^2)A_1 + \tfrac{3}{4}b(A_1^3 + A_1^2 A_3 + 2A_1 A_3^2) &= 0, \\
(a - 9\omega^2)A_3 + \tfrac{1}{4}b(A_1^3 + 6A_1^2 A_3 + 3A_3^3) &= f.
\end{aligned} \tag{V.6-3}$$

Note that (V.6-1) represents a $\tfrac{1}{3}$ subharmonic since the fundamental frequency is $\tfrac{1}{3}$ the frequency of the forcing function to the differential equation (V.6-2). The existence of subharmonics has been experimentally verified (see, e.g., Hayashi). Our objective is to obtain mathematical verification along with explicit bounds on the difference between approximate and exact subharmonic solutions.

The approach used so far in this chapter is easily adaptable to the problem of determining whether there is an exact solution near the subharmonic predicted by harmonic balance. First redefine $\tilde{L}$ as follows (compare with (V.2-16):

$$\begin{aligned}
(Lx)(t) &\sim \frac{a_0}{2} + \sum_{k=1}^{\infty} (a_k \cos k\omega t + b_k \sin k\omega t) \\
(\tilde{L}x)(t) &= \sum_{k=1,3} (a_k \cos k\omega t + b_k \sin k\omega t).
\end{aligned} \tag{V.6-4}$$

† For each frequency, there is a different Banach space.

If $a_0 = a_2 = b_2 = 0$, then $\tilde{L}x = P_3 Lx$.[†] It then easily follows that the use of Theorem II.3.2 (using the same arguments as in Sec. V.3 and V.4) leads to finding a $\gamma \in [0, 1)$ satisfying

$$3bcT\left(|A_1| + |A_3| + \frac{cTb[\frac{3}{4}|A_1^2 A_3| + A_1 A_3^2| + \frac{3}{4}|A_1 A_3^2| + \frac{1}{4}|A_3^3|]}{1 - \gamma}\right)^2 \le \gamma$$

$$(V.6\text{-}5)$$

with c given by (V.4-5). If there is such a γ, then there is a fixed point x^* satisfying

$$\|x^* - x_0\| \le \frac{cTb[\frac{3}{4}|A_1^2 A_3 + A_1 A_3^2| + \frac{3}{4}|A_1 A_3^2| + \frac{1}{4}|A_3^3|]}{1 - \gamma}$$

$$(V.6\text{-}6)$$

with $\|x\| = \max_t |x(t)|$ and x_0 is the harmonic balance approximation satifying (V.6-1) and (V.6-3).

The established existence of the fixed point near the approximation does not yet prove the existence of a subharmonic. It must be shown that $A_1^* \neq 0$ with A_1^* the coefficient of the $\cos \omega t$ term in the Fourier series of $x^*(t)$ (which equals the sum of its Fourier series). This can be done as follows. If $\max_t |x^*(t) - x_0(t)| \le e$, then it easily follows from

$$A_1^* - A_1 = 2T^{-1} \int_0^T \left[x^*(t) - x_0(t)\right] \cos \omega t \, dt$$

$$(V.6\text{-}7)$$

that $|A_1^* - A_1| \le (4/\pi)e$. Thus, $(4/\pi)e < A_1$ implies that $A_1^* \neq 0$.

Let us consider the following numerical case:

$$a = 2, \qquad b = 0.05, \qquad f = 2$$
$$A_1 = 1.0, \qquad A_3 = -0.122, \qquad \omega = 4.28 \ (T = 1.47).$$

$$(V.6\text{-}8)$$

For this case, $c = 0.41$ and it is found that $\gamma = 0.12$ satisfies (V.6-5) and also that

$$\max_t |x^*(t) - x_0(t)| \le 0.0031.$$

$$(V.6\text{-}9)$$

Furthermore, $A_1^* \neq 0$ so that the existence of a subharmonic is proved.

Remark. The mathematically proved existence of a periodic solution does not necessarily mean that such a periodic solution will be witnessed in a physical

[†] With $a_0 = a_2 = b_2 = 0$, we also have from (V.A-3),
$$(P_3 Lx)(t) = \frac{1}{T} \int_0^T \frac{\sin 7/2 \, \omega u}{\sin \omega u/2} (Lx)(t + u) \, du.$$

system described by the equations. For example, the subharmonic may be excited only with certain initial conditions and, further, if the subharmonic is not stable it may not be maintained under ever present perturbations. We refer the reader to Hayashi for discussion and experimental results. This is an area where further mathematical analysis would be useful.

V.7 TIME LAG SYSTEMS

The introduction of time lags into the system does not substantially complicate the contraction mapping analysis. For example, consider the system represented by

$$x = LN(D_h x) \tag{V.7-1}$$

where $y = D_h x$ is defined by

$$y(t) = x(t - h), \qquad h > 0. \tag{V.7-2}$$

If the Banach space is C_T (assume scalar functions for simplicity), then

$$\|D_h\| = 1. \tag{V.7-3}$$

This follows easily from

$$
\begin{aligned}
\|D_h\| &= \sup \left\{ \|D_h x\| : x \in C_T, \|x\| = 1 \right\} \\
&= \sup \left\{ \max_t |x(t - h)| : x \in C_T, \max_t |x(t)| = 1 \right\}.
\end{aligned}
\tag{V.7-4}
$$

Assume that N maps C_T into itself. If N is differentiable (i.e., has a Fréchet derivative) at x_0 then $N(D_h x)$ has a derivative at x_0 given by $N'(D_h x_0) D_h$ [see (I.5-5)]. Then $LN(D_h)$ has a derivative at x_0 given by $LN'(D_h x_0) D_h$ [see (I.5-6)]. The norm is easily bounded:

$$\|LN'(D_h x_0) D_h\| \leqq \|L\| \cdot \|N'(D_h x_0)\|. \tag{V.7-5}$$

Conceptually, the introduction of the time delay offers no great difficulty as compared to the case without the time delay. However, it will generally complicate the arithmetic involved in examples, in particular, in obtaining the solution to $x_0 = \tilde{L}N(x_0)$. This relative lack of complication going from differential equations to difference-differential equations is not typical. In existence, uniqueness, and stability considerations one must consider initial function conditions in difference-differential equations while the initial conditions for differential equations are merely at one time.

In view of the above remarks, consider the following difference-differential equation:

$$\ddot{y}(t) + ay(t) + by_h^3(t) = f\cos\omega t, \tag{V.7-6}$$

$$y_h(t) = y(t - h) . \tag{V.7-7}$$

This is Duffing's equation but with the argument of the cubic term retarded. The corresponding operator equation is

$$x = LN(x) = L[N_1(D_h x) + F], \tag{V.7-8}$$

where $y = N_1(x)$ is defined by the cubic nonlinearity, L is the same linear operator as in Sec. V.4 and F represents the forcing function $f\cos\omega t$. It is clear that

$$\|LN'(x)\| \leq \|L\| \cdot \|N_1'(D_h x)\| \tag{V.7-9}$$

and that the analysis will be completely analogous to that of Sec. V.3 or Sec. V.4 except that the approximate solution, x_0, will be different. Note that the Banach space is the space of continuous periodic functions, not the space of functions continuous on one period.

To obtain the equivalent linearization approximation let

$$y(t) = A\cos\omega t + B\sin\omega t$$
$$= C\sin(\omega t + \theta), \tag{V.7-10}$$

where

$$C = \sqrt{A^2 + B^2}, \qquad \theta = \tan^{-1}\left(\frac{A}{B}\right). \tag{V.7-11}$$

Substitution of this function into (V.7-6) yields

$$-\omega^2 C\sin(\omega t + \theta) + aC\sin(\omega t + \theta) + \tfrac{3}{4}C^3 b[\cos\omega h\sin(\omega t + \theta)$$
$$- \sin\omega h\cos(\omega t + \theta) + \text{third harmonics}] = f\cos\omega t. \tag{V.7-12}$$

The harmonic balance approximation is obtained by neglecting the third harmonics and equating coefficients of $\cos\omega t$ and $\sin\omega t$. It is interesting to compare the approximate solution obtained for the difference-differential equation with that obtained for the following differential equation (Duffing's equation with a damping term):

$$\ddot{y}(t) + d\dot{y}(t) + ay(t) + by^3(t) = f\cos\omega t. \tag{V.7-13}$$

Substituting (V.7-10) into (V.7-13) yields

$$-\omega^2 C \sin(\omega t + \theta) + d\omega C \cos(\omega t + \theta) + aC \sin(\omega t + \theta)$$
$$+ \tfrac{3}{4} bC^3 \sin(\omega t + \theta) + \text{third harmonics} = f \cos \omega t. \tag{V.7-14}$$

Comparing (V.7-12) and (V.7-14), it is seen that, as far as harmonic balance (or equivalent linearization) is concerned, the effect of the lag is to introduce a damping term with damping coefficient d,

$$d = -(\sin \omega h)\tfrac{3}{4} C^2 b/\omega. \tag{V.7-15}$$

(Also, one other term is multiplied by $\cos \omega h$.)

V.8 AUTONOMOUS SYSTEMS

Autonomous differential equations are those whose right-hand sides do not depend on time. The describing function method is used often by control engineers for the prediction of self-oscillations of systems described by autonomous differential equations. It would seem at first glance that our approach should be appropriate for analysis of this problem. Suppose the describing function method indicates that there exists a nontrivial periodic solution x_0 to the operator equation

$$x_0 = \tilde{L} N(x_0) \tag{V.8-1}$$

with $\tilde{L}$ defined by (V.2-16). Usually, $N(0) = 0$, so that it is of interest to investigate whether there is any nontrivial solution to the exact equation near x_0. If our method is successful, then we can guarantee that the fixed point is not the trivial solution ($x = 0$) if

$$\frac{\|LN(x_0) - x_0\|}{1 - \gamma} < \|x_0\| \tag{V.8-2}$$

since the fixed point x^* satisfies

$$\|x^* - x_0\| \leq \frac{\|LN(x_0) - x_0\|}{1 - \gamma}. \tag{V.8-3}$$

Let us discuss some difficulties in applying our contraction mapping approach. First of all, we may have no a priori information on the period of the oscillation; it could be different from that of the periodic solution predicted by describing functions. With nonautonomous periodic systems, the period of the periodic solution will usually be related to the periodicity of the system but autonomous systems are periodic of every period. This consideration is not necessarily a serious stumbling block since one may normalize the time parameter and bring out the period as a parameter. What prevents our approach from being successful is the noncontractive nature of fixed points associated with nontrivial autonomous oscillations. This can be seen most simply by recognizing that if $x(t)$ is a periodic solution, then so is $x(t + \epsilon)$ for any ϵ. Hence, one could not have the uniqueness associated with $x(t)$ that would be present if $x(t)$ corresponded to a fixed point of a contraction mapping.

The noncontractive nature can also be seen by considering the vector differential equation

$$\dot{x}(t) = Ax(t) + n(x(t)) \tag{V.8-4}$$

with A a constant real-valued matrix and $n(u)$ a real-valued function having continuous partial derivatives with respect to all of the components of the vector u. Suppose there is a continuous T-periodic x^* satisfying (V.8-4). The equation of first variation corresponding to (V.8-4) and x^* is

$$\dot{y}(t) = Ay(t) + \left[\frac{\partial n}{\partial x^*}\right]y(t), \tag{V.8-5}$$

where $[\partial n/\partial x^*]$ is the (time-varying) Jacobian matrix with entries $\partial n_i/\partial x_j$ evaluated along the solution defined by x^*. By differentiation of (V.8-4) with respect to time, it is seen that $y(t) = \dot{x}^*(t)$ satisfies (V.8-5) and hence the equivalent integral equation (assuming $\dot{x} = Ax$ has no nontrivial T-periodic solutions)

$$\dot{x}^*(t) = \int_0^T G(t, s)\left[\frac{\partial n}{\partial x^*}\right]\dot{x}^*(s)\, ds. \tag{V.8-6}$$

The linear integral operator defined by the right-hand side of (V.8-6) is the derivative of LN where $x = LNx$ represents the nonlinear integral equation equivalent to (V.8-4). Hence

$$\|\dot{x}^*\| \leq \|LN'(x^*)\| \cdot \|\dot{x}^*\|. \tag{V.8-7}$$

If $\|LN'(x^*)\| < 1$, then $\dot{x}^*$ must be zero and there cannot be a nonconstant periodic solution.

Since a straightforward application of the CMT fails, the following can be done. The CMT is applied in a subspace of periodic functions defined by the exclusion of the fundamental frequency of the periodic solution. The fundamental component is then handled by a separate argument.

It is convenient to consider the system described in Sec. IV.2 with $r = 0$, with the linear operator L described by (IV.2-8) and with $h \in L_1(-\infty, \infty)$. Further, assume the nonlinear function satisfies $n(0) = 0$ and $\beta > -\alpha$ in addition to (IV.2-3). We shall use the space B_T but it is important to bear in mind that $T(= 2\pi/\omega)$ is not to be considered fixed throughout the discussion since the period of the actual oscillation (if one exists) need not be the same as that of the describing function approximation.

If $x \in B_T$ and

$$x = \lim_{M \to \infty} \sum_{m=-M}^{M} c_m e^{im\omega t} \quad (c_{-m} = \bar{c}_m) \tag{V.8-8}$$

(the limit taken in the sense of the B_T norm), define

$$u_k(c_0, c_1, c_2, \ldots) = T^{-1} \int_0^T n(x(t)) e^{-ik\omega t} dt, \tag{V.8-9}$$

$$u(c_1) = u_1(0, c_1, 0, 0, \ldots).$$

We assume, without loss of generality, that c_1 is real. Note that $u(c_1)/c_1$ is a describing function.† Also recall, from the second footnote of Sec. IV.4, that $u(c_1)/c_1$ is real (although shown in a different notation in that footnote). For x to be a B_T function satisfying $x = -LNx$ we must clearly have the following balance of Fourier coefficients:

$$c_k = -H(ik\omega)u_k(c_0, c_1, c_2, \ldots), \qquad k = 0, 1, 2, \ldots. \tag{V.8-10}$$

It is assumed that we have already found a describing function approximation x_0 of frequency ω_0 satisfying

$$x_0(t) = 2a \cos \omega_0 t = a(e^{i\omega_0 t} + e^{-i\omega_0 t}), \qquad a > 0, \tag{V.8-11}$$

$$a = -H(i\omega_0)u(a). \tag{V.8-12}$$

We wish to determine conditions for the existence of a periodic function satisfying (V.8-10) which is close in amplitude and frequency to that of x_0. To do so, let us first break up (V.8-10) as follows:

$$-c_1 H^{-1}(i\omega) - u(c_1) = u_1(c_0, c_1, c_2, \ldots) - u(c_1), \tag{V.8-13}$$

$$c_k = -H(ik\omega)u_k(c_0, c_1, c_2, \ldots), \quad k = 0, 2, 3, \ldots. \tag{V.8-14}$$

The first equation is merely the $k = 1$ term of (V.8-10) rewritten with a term subtracted from both sides.

†

$$\frac{u(c_1)}{c_1} = \frac{T^{-1} \int_0^T n(2c_1 \cos \omega t) e^{-i\omega t} dt}{2c_1}$$

$$= \frac{2T^{-1} \int_0^T n(2c_1 \cos \omega t) e^{-i\omega t} dt}{2c_1} = N(2c_1)$$

with N here representing a describing function [see (IV.1-1) and the discussion below (IV. 4-6)].

Assume in the following that c_1 and ω are real numbers in a rectangle D defined by $|\omega - \omega_0| \leq \Delta$, $|c_1 - a| \leq \delta$. Let us apply the CMT to find a fixed point of (V.8-14). We work in B_T', the subspace of B_T, which omits the fundamental frequency, i.e., the space consisting of the closure of the set of real-valued linear combinations of $\{\ldots, e^{-i3\omega t}, e^{-i2\omega t}, 1, e^{i2\omega t}, e^{i3\omega t}, \ldots\}$. This space may be seen to be a closed subspace of B_T and hence a Banach space by itself.

If $x' \in B_T'$, then (with the limit in the B_T' norm)

$$x' = \lim_{\substack{M \to \infty}} \sum_{\substack{m=-M \\ m \neq \pm 1}}^{M} c_m e^{im\omega t} \tag{V.8-15}$$

and

$$\|x'\| = \left(\sum_{\substack{m=-\infty \\ m \neq \pm 1}}^{\infty} |c_m|^2 \right)^{1/2}. \tag{V.8-16}$$

Letting F represent the operator defined by the right-hand side of (V.8-14), it is easily verified that if $x_1', x_2' \in B_T'$,

$$\|Fx_1' - Fx_2'\| \leq \sup_{m=0,2,3,\ldots} |H(im\omega)| \, \beta \, \|x_1' - x_2'\|. \tag{V.8-17}$$

If

$$\sup_{|\omega - \omega_0| \leq \Delta} \sup_{m=0,2,3,\ldots} |H(im\omega)| \beta = \gamma < 1, \tag{V.8-18}$$

then for each $(\omega, c_1) \in D$, there is a function x' in B_T' having Fourier coefficients c_k satisfying (V.8-14). From (II.1-8) (taking that $x_0' = 0$) and (V.8-16),

$$\|x'\| = \left(\sum_{\substack{k=-\infty \\ k \neq \pm 1}}^{\infty} |c_k|^2 \right)^{1/2} \leq \frac{\sqrt{2}\,\gamma(a + \delta)}{1 - \gamma}, \tag{V.8-19}†$$

which is satisfied for all $(\omega, c_1) \in D$.

For each $(\omega, c_1) \in D$, (V.8-14) is satisfied with the above $c_k(\omega, c_1)$ for $k = 0$, $\pm 2, \pm 3, \ldots$. We substitute these $c_k(\omega, c_1)$ into (V.8-13) and we have then an equation depending on the two variables ω and c_1.

It remains to show that there is a pair $(\omega, c_1) \in D$ such that (V.8-13) can be satisfied. To do this we use a result based on the theory of rotation of vector fields. We will not go into this theory here because the background necessary

$$
† \qquad \|Fx_0' - x_0'\|^2 = \|Fx_0'\|^2 = \sum_{k \neq \pm 1} |H(ik\omega)u_k(0, c_1, 0, 0, \ldots)|^2
$$
$$
\leq \sup_{k \neq \pm 1} |H(ik\omega)|^2 \, T^{-1} \int_0^T [n(2c_1 \cos \omega t)]^2 \, dt
$$
$$
\leq \sup_{k \neq \pm 1} |H(ik\omega)|^2 2\beta^2 c_1^2.
$$

to adequately explain it (not just throwing out definitions and theorems) would take too much space; instead, we refer the interested reader to a most readable book by Krasnosel'skiy *et al.* The reader can also use this book as an introduction to fixed point theorems not of the contractive type (in particular, Brouwer's fixed point theorem).

The result we need is the following.

Theorem V.8.1

Let Φ and θ be continuous maps of D, a rectangle in the plane, into the plane. That is, for $(u, v) \in D$, $\Phi(u, v) = (\Phi_1(u, v), \Phi_2(u, v))$. Suppose that for some unique (u_0, v_0) in D, $\Phi(u_0, v_0) = (0, 0)$ and that the Jacobian

$$\left|\frac{\partial \Phi(u, v)}{\partial(u, v)}\right|_{\substack{u=u_0 \\ v=v_0}} \equiv \begin{vmatrix} \dfrac{\partial \Phi_1(u, v)}{\partial u} & \dfrac{\partial \Phi_1(u, v)}{\partial v} \\ \dfrac{\partial \Phi_2(u, v)}{\partial u} & \dfrac{\partial \Phi_2(u, v)}{\partial v} \end{vmatrix}_{\substack{u=u_0 \\ v=v_0}} \neq 0. \qquad \text{(V.8-20)}$$

Further suppose that

$$|\Phi(u, v)| > |\theta(u, v)| \qquad \text{(V.8-21)}$$

for all $(u, v) \in \partial D$ where ∂D is the boundary of D. Then $\Phi(u, v) - \theta(u, v)$ has a zero in D. In other words, there is a $(u, v) \in D$ such that

$$\Phi(u, v) = \theta(u, v). \qquad \text{(V.8-22)}$$

Remark. As already mentioned, this theorem is based on the theory of rotation of vector fields which is covered well in Krasnosel'skiy *et al.* We mention, in passing, a simpler result than needed but which contains the main ingredient of the more general result. In particular we quote Rouché's theorem which can be found in most books on complex variables (e.g., Titchmarch [2], p. 116, or Evgrafov, p. 99):

Rouche's Theorem. Let $f(z)$ and $g(z)$ be analytic in a domain D and continuous up to its boundary ∂D. If $|f(z)| > |g(z)|$ on ∂D, then $f(z)$ and $f(z) + g(z)$ have the same number of zeros in D.

The proof of Rouché's theorem is based on the principle of the argument with which the reader should be familiar in connection with Nyquist's criterion. Incidentally, the argument principle is itself derivable using the theory of rotation of vector fields.

Now to the problem of verifying the existence of a (ω, c_1) in the rectangle D so that (V.8-13) is satisfied.

To use Theorem V.8.1 in verifying the existence of $(\omega, c_1) \in D$ for satisfaction of (V.8-13), we identify the left-hand side of (V.8-13) with Φ by considering the real and imaginary parts in the following manner:

$$\Phi(\omega, c_1) = (-c_1 \operatorname{Re} H^{-1}(i\omega) - u(c_1), -c_1 \operatorname{Im} H^{-1}(i\omega)) \qquad \text{(V.8-23)}$$

where we have used the realness of the describing function $u(c_1)/c_1$ and of c_1. The continuity of Φ on D can be verified using the continuity of Fourier transforms of $L_1(-\infty, \infty)$ functions if $H(i\omega) \neq 0$ for $|\omega - \omega_0| \leq \Delta$. The function θ is defined in terms of the right-hand side of (V.8-13):

$$\theta(\omega, c_1) = (\operatorname{Re} u_1(c_0, c_1, c_2, \dots) - u(c_1), \operatorname{Im} u_1(c_0, c_1, c_2, \dots)). \qquad \text{(V.8-24)}$$

The continuity of θ may be verified by using Theorem II.1.4 to imply the continuous dependence of the fixed points in B'_T on (ω, c_1). To satisfy the conditions of Theorem V.8.1, it is assumed that $\Phi(\omega, c_1) = 0$ has (ω_0, a) as its only solution in D [see (V.8-12)]. The nonvanishing of the Jacobian reduces to

$$\left| \frac{\partial \Phi(\omega, c_1)}{\partial(\omega, c_1)} \right|_{\substack{\omega = \omega_0 \\ c_1 = a}} = \left[u(a) - a \frac{d}{da} u(a) \right] \frac{d}{d\omega} \operatorname{Im} H^{-1}(i\omega) \Big|_{\omega = \omega_0} \neq 0 \qquad \text{(V.8-25)}$$

using (V.8-12) which implies $\operatorname{Im} H^{-1}(i\omega_0) = 0$ since a and $u(a)$ are real. Next, note that for all $(\omega, c_1) \in D$,

$$
\begin{aligned}
|\theta(\omega, c_1)|^2 &= \left| T^{-1} \int_0^T [n(x(t)) - n(2c_1 \cos \omega t)] e^{-i\omega t} \, dt \right|^2 \\
&\leq \beta^2 T^{-2} \int_0^T [x(t) - 2c_1 \cos \omega t]^2 \, dt \int_0^T |e^{-i\omega t}|^2 \, dt \\
&= \beta^2 \|x'\|^2 \\
&\leq \beta^2 \left[\frac{\sqrt{2}\, \gamma(a + \delta)}{1 - \gamma} \right]^2.
\end{aligned}
\qquad \text{(V.8-26)}
$$

The last step used (V.8-19) assuming x' is the fixed point in B'_T already found using the CMT. Hence, if

$$\min_{(\omega, c_1) \in \partial D} |\Phi(\omega, c_1)| > \sqrt{2}\, \beta \frac{\gamma(a + \delta)}{1 - \gamma} \qquad \text{(V.8-27)}$$

we can use Theorem V.8.1 to guarantee the existence of a $(\omega, c_1) \in D$ for satisfaction of (V.8-13) and thus a periodic solution to the autonomous system. Let us summarize the key conditions:

(i) $H(i\omega) \neq 0$ for $|\omega - \omega_0| \leq \Delta$,   (V.8-28)

(ii) $\Phi(\omega, c_1) \neq 0$ for any $(\omega, c_1) \in D$ other than (ω_0, a),   (V.8-29)

(iii) $\left[u(a) - a\dfrac{d}{da}u(a)\right]\dfrac{d}{d\omega} \operatorname{Im} H^{-1}(i\omega)\Big|_{\omega=\omega_0} \neq 0,$ (V.8-30)

(iv) $\beta \max\left\{|H(0)|, \sup\limits_{\omega>2(\omega_0-\Delta)}|H(i\omega)|\right\}$

$$< \frac{\min\limits_{(\omega,c_1)\in\partial D}|\Phi(\omega, c_1)|}{\min\limits_{(\omega,c_1)\in\partial D}|\Phi(\omega, c_1)| + \sqrt{2}\,\beta(a + \delta)}.$$ (V.8-31)

Conditions (V.8-18) and (V.8-27) are simultaneously satisfied with the satisfaction of the last inequality. Also, note that $[u(a) - a\,du(a)/da]$ is $(-a^2)$ times the derivative of the describing function $u(a)/a$. Under these conditions, there is a periodic solution to $x = -LNx$ of frequency ω with $|\omega - \omega_0| \leq \Delta$, and with Fourier coefficients satisfying $|c_1 - a| \leq \delta$ and

$$\left(\sum_{\substack{k=-\infty \\ k\neq\pm 1}}^{\infty} |c_k|^2\right)^{1/2} \leq \beta^{-1} \min_{(\omega,c_1)\in\partial D}|\Phi(\omega, c_1)|.$$ (V.8-32)

Condition (V.8-32) follows from (V.8-19) with γ bounded by the right-hand side of (V.8-31).

We can try to improve the preceding result in two ways. First of all, note that we have just used the Lipschitz constant β and we have not taken advantage of knowledge of α [see (IV.2-3)]. As in Sec. III.1, we can manipulate the equation $x = -LNx$ into the form

$$x = -(I + \tfrac{1}{2}(\alpha + \beta)L)^{-1}L\tilde{N}x$$ (V.8-33)

assuming the inverse exists. Then condition (V.8-31) becomes

$$\frac{1}{2}(\beta - \alpha) \max\left\{|\tilde{H}(0)|, \sup_{\omega>2(\omega_0-\Delta)}|\tilde{H}(i\omega)|\right\}$$

$$< \frac{\min\limits_{(\omega,c_1\in)\partial D}|\Phi(\omega, c_1)|}{\min\limits_{(\omega,c_1)\in\partial D}|\Phi(\omega, c_1)| + \sqrt{2}\,\beta(a + \delta)}$$ (V.8-34)

with

$$\tilde{H}(i\omega) = \frac{H(i\omega)}{1 + \tfrac{1}{2}(\alpha + \beta)H(i\omega)}, \qquad H(i\omega) \neq -\frac{2}{\alpha + \beta}.$$ (V.8-35)

The second part of (V.8-35) guarantees the existence of the inverse mentioned above.

It would be more pleasing if the left-hand sides of (V.8-31) and (V.8-34) did not depend on $H(0)$ since $|H(i\omega)|$ often takes on its maximum value at $\omega = 0$. We can remove this dependence if n is an odd function, $n(-u) = -n(u)$. In this case, we can work in the following subspace of B_T,

$$B_T^0 = \left\{ x: x \in B_T, \int_0^T x(t)e^{-ik\omega t}\, dt = 0 \qquad \text{for } k \text{ even} \right\}. \tag{V.8-36}$$

Functions in B_T^0 satisfy $x(t) = -x(t + \tfrac{1}{2}T)$. The odd nonlinearity preserves this property and hence maps B_T^0 into itself. Working in the space B_T^0 requires that (V.8-14) be satisfied only for $k = \pm 3, \pm 5, \ldots$ instead of $k = 0, \pm 2, \pm 3, \ldots$. The CMT analysis is applied in $B_T^0 \cap B_T'$ which is a Banach space by itself. It is then easily seen that condition (V.8-34) reduces to

$$\tfrac{1}{2}(\beta - \alpha) \sup_{\omega > 3(\omega_0 - \Delta)} |\tilde{H}(i\omega)| < \frac{\min\limits_{(\omega, c_1) \in \partial D} |\Phi(\omega, c_1)|}{\min\limits_{(\omega, c_1) \in \partial D} |\Phi(\omega, c_1)| + \sqrt{2}\,\beta(a + \delta)}. \tag{V.8-37}$$

The left-hand side of (V.8-37) shows the justification of the describing function method to depend on the departure of the nonlinearity from linearity, measured by $(\beta - \alpha)$, and the low pass filtering effect, measured in this case by

$$\sup_{\omega > 3(\omega_0 - \Delta)} |\tilde{H}(i\omega)|.$$

Exercise V.8.1

What happens to the error bound of (V.8-32) as we go from (V.8-31) to (V.8-34) to (V.8-37)?

Example V.8.1

Let

$$H(i\omega) = \frac{1}{(i\omega + 1)^4}, \tag{V.8-38}$$

and let n be the saturation function

$$\begin{aligned} n(\sigma) &= 5\sigma & |\sigma| &\leq 1, \\ &= 5\,\text{sgn}\,\sigma & |\sigma| &> 1. \end{aligned} \tag{V.8-39}$$

For this n, $\beta = 5$, $\alpha = 0$, and (referring to the first footnote in this section)

$$\begin{aligned} u(c) &= 5c & 0 &\leq c \leq \tfrac{1}{2}, \\ &= \frac{10c}{\pi}\left[\sin^{-1}\frac{1}{2c} + \frac{1}{2c}\sqrt{1 - (2c)^{-2}} \right] & c &> \tfrac{1}{2}, \end{aligned} \tag{V.8-40}$$

as can easily be verified directly or by referring to a table of describing functions (e.g., Gelb and Vander Velde, p. 58, or Hsu and Meyer, p. 199) since

$u(c)/c$ is a describing function. The describing function approximation x_0, from (V.8-11) and (V.8-12), is described by $\omega_0 = 1$ (the frequency at which $\underline{/H(i\omega)}$ $= -180°$), $u(a)/a = 4$ and $a = 0.728$.

The verification of (V.8-37) can be checked in different ways, depending on how much of it is done with a computer. For this example, the following was done. First, (V.8-37) was manipulated so that

$$\min_{(\omega,\, c_1) \in \partial D} |\Phi(\omega, c_1)| > f(\Delta, \delta), \tag{V.8-41}$$

the right-hand side being a function of Δ and δ. Values of $f(\Delta, \delta)$ were computed and printed out for ranges of Δ and δ between 0 and Δ_{max} and δ_{max}, respectively. $|\Phi(\omega, c_1)|$, evaluated for $|\omega - \omega_0| \leq \Delta_{max}$ and $|c_1 - a| \leq \delta_{max}$, was computed and printed out in a rectangular array representing the (ω, c_1) plane. Then it is a matter of inspecting the minimum value of $|\Phi(\omega, c_1)|$ on boundaries of different rectangles and comparing them with the corresponding $f(\Delta, \delta)$ to see if (V.8-41) is satisfied. Best estimates of the frequency and amplitude of the fundamental are obtained with the smallest rectangles. Satisfaction of (V.8-41) was found for

$$\Delta = 0.016, \qquad \delta = 0.035$$

with a corresponding

$$\min_{(\omega,\, c_1) \in \partial D} |\Phi(\omega, c_1)| = 0.0784.$$

We must also see that conditions (V.8-28)–(V.8-30) are satisfied, but this is easily verified.

We finally note that the above approach of the using the CMT in a subspace excluding the fundamental and then using a separate argument for the fundamental can also be used for nonautonomous systems. It is more complicated than a straightforward application of the CMT but may be applicable in some cases when the straightforward use of the CMT is not. In the autonomous case, the theory of rotation of vector fields is used in the plane since there are two unknowns: the amplitude and frequency of the fundamental (the phase being unimportant). In the nonautonomous case, the frequency, but not the phase, may be assumed known so that the plane is still used.

V.9 NOTES

Linear vector differential equations are discussed in many places, e.g., Hochstadt, Struble, Zadeh and Desoer. The derivation of the equivalent integral equation in Sec. V.2 follows Halanay, Chapter 3. The harmonic balance analysis of Duffing's equation is given in Stoker and McLachlan. Duffing's equation is discussed from a feedback point of view in Graham and McRuer. The contraction mapping analyses are based on Holtzman [4] and [1]. The arguments leading to (V.8-31) are based on Kudrewicz which also has generalizations. Time-lag systems are discussed from different points of view in Halanay and Minorsky.

The subject of best trigonometric polynomials in C_T discussed in the Appendix is a much studied question in the field of mathematics called approximation theory. The reader can consult the recent books by Cheney and Lorentz which also have further references.

APPENDIX

Supplement on Trigonometric Polynomials

For simplicity, let C_T be the space of continuous real-valued T-periodic scalar functions.

It is of interest to relate Fourier series partial sums to best trigonometric polynomial approximations in C_T. If $x \in C_T$, let $T_n x$ be a trigonometric polynomial of degree $\leq n$ which satisfies for any y, a trigonometric polynomial of degree $\leq n$,

$$\|x - T_n x\| \leq \|x - y\|. \tag{V.A-1}$$

This T_n (which is a nonlinear operator whose existence may be proved) gives the best trigonometric polynomial approximation to x (best in the C_T norm). Using the triangle inequality and the projection property of P_n which implies $T_n x = P_n T_n x$,

$$\|x - P_n x\| \leq \|x - T_n x\| + \|P_n(x - T_n x)\|$$
$$\leq \|x - T_n x\|(1 + \|P_n\|). \tag{V.A-2}$$

For (V.A-2) to be helpful, we need an upper bound on $\|P_n\|$; a lower bound was given in (V.1-9). We have from (V.1-5), (IV.1-1), and (IV.1-2)

$$(P_n x)(t) = \frac{1}{T} \int_0^T x(s) \left[1 + 2 \sum_{k=1}^n (\cos k\omega s \cos k\omega t + \sin k\omega s \sin k\omega t) \right] ds$$

$$= \frac{1}{T} \int_0^T x(s) \left[1 + 2 \sum_{k=1}^n \cos k\omega(s - t) \right] ds$$

$$= \frac{1}{T} \int_0^T x(s) \frac{\sin \left[(n + \frac{1}{2})\omega(s - t) \right]}{\sin \omega(s - t)/2} ds$$

$$= \frac{1}{T} \int_0^T x(t + u) \frac{\sin (n + \frac{1}{2})\omega u}{\sin (\omega u/2)} du \qquad \text{(V.A-3)†}$$

which shows that P_n may be represented by a linear integral operator (in fact, a convolution type operation). It then follows from the definition of the norm of a linear operator and the change of variable $\theta = \omega u/2$ that

$$\|P_n\| \leq \frac{2}{\pi} \int_0^{\pi/2} \left| \frac{\sin (2n + 1)\theta}{\sin \theta} \right| d\theta. \qquad \text{(V.A-4)}$$

It may be shown that the inequality of (V.A-4) may be replaced by equality (Lorentz, p. 5). As n gets large, the integral in (V.A-4) becomes quite complicated. It can easily be bounded by using the following two inequalities,

$$|\sin m\theta| \leq m |\sin \theta| \qquad (m \text{ an integer})$$

$$\sin \theta \geq \frac{2}{\pi}\theta. \qquad 0 \leq \theta \leq \pi/2 \qquad \text{(V.A-5)}$$

to obtain

$$\|P_n\| \leq \frac{2}{\pi} \int_0^\lambda (2n + 1) \, d\theta + \int_\lambda^{\pi/2} \frac{d\theta}{\theta} \qquad (0 < \lambda < \pi/2)$$

$$= \frac{2}{\pi} (2n + 1)\lambda + \ln \left(\frac{\pi}{2\lambda} \right). \qquad \text{(V.A-6)}$$

Letting $\lambda = \pi/[2(2n + 1)]$ (which makes the derivative of the last expression vanish) yields

$$\|P_n\| \leq 1 + \ln (2n + 1). \qquad \text{(V.A-7)}$$

From (V.A-2) and (V.A-7),

$$\|x - P_n x\| \leq [2 + \ln (2n + 1)]\|x - T_n x\|. \qquad \text{(V.A-8)}$$

† Using the identity $2 \cos \alpha \sin \beta = \sin (\alpha + \beta) - \sin (\alpha - \beta)$,

$$\left[1 + 2 \sum_{k=1}^n \cos k\theta \right] \sin \frac{\theta}{2} = \sin \frac{\theta}{2} + \sum_{k=1}^n \left[\sin (k + \tfrac{1}{2})\theta - \sin (k - \tfrac{1}{2})\theta \right]$$

$$= \sin (n + \tfrac{1}{2})\theta .$$

For $n = 1$, (V.A-8) gives $\|x - P_1 x\| \leq 3.1 \|x - T_1 x\|$. With $n = 1$, the exact value of the integral in (V.A-4) is also easily evaluated to be 1.43 so that $\|x - P_1 x\| \leq 2.43 \|x - T_1 x\|$.

We can use (V.A-8) not only to compare using P_n with using T_n for approximation but also to bound the rate at which $P_n x \to x$ for periodic functions with continuous derivatives. It may be shown that

$$\|x - T_n x\| \leq \frac{T}{4(n + 1)} \|\dot{x}\| \qquad \text{(V.A-9)}$$

(Cheney, p. 142) so that

$$\|x - P_n x\| \leq \frac{T[2 + \ln(2n + 1)]}{4(n + 1)} \|\dot{x}\| . \qquad \text{(V.A-10)}$$

Condition (V.A-10) is a good bound for showing asymptotic behavior as $n \to \infty$ but for some specific values of n, sharper estimates are available. For example, with $n = 1$ use the exact value of $\|P_n\|$ in (V.A-2).

The purpose of the above discussion is to compare the use of P_n with the best trigonometric polynomial approximation in C_T since P_n is often used to obtain approximate periodic solutions of nonlinear systems and the C_T norm is of practical interest. We are not aware of the use of any trigonometric polynomials other than $P_n x$ to obtain approximate periodic solutions. The use of T_n is much more complicated than P_n. There are also linear operators simpler to use than T_n but with C_T norm properties better in some respects than P_n. In particular, there is a linear operator U_n mapping periodic functions with continuous derivatives into trigonometric polynomials of degree $\leq n$ which satisfies

$$\|x - U_n x\| \leq \frac{T}{4(n + 1)} \|\dot{x}\| \qquad \text{(V.A-11)}$$

which is the same bound as obtained for the best trigonometric polynomial approximation, (V.A-9) (Cheney, pp. 140–142). At first glance, this may seem to contradict our earlier statement that T_n is a nonlinear operator. However, given a particular x, we could have $\|x - T_n x\| < \|x - U_n x\|$. What (V.A-9) and (V.A-11) show is that T_n and U_n give the same uniform bound (worst case bound) over the class of x's with continuous derivatives. In fact, $U_n x$ could be a worse approximation to a particular x than is $P_n x$ if x is a trigonometric polynomial of degree $\leq n$; U_n is not a projection operator.

If V_n is a bounded linear projection of C_T onto the subspace of trigonometric polynomials of degree $\leq n$, then $\|P_n\| \leq \|V_n\|$ (Cheney, p. 212). Hence, the use of P_n as compared to all projections V_n provides the smallest bound corresponding to (V.A-2).

The last comment incidentally raises the question of uniqueness of projections. We defined projections in Sec. IV.1 as being the linear mappings taking elements in a linear space into their (unique) components in subspaces which direct sum decompose the space. Given the direct sum decomposition, the projection operators are uniquely specified. However, if we are given a subspace there is not another unique subspace that yields a direct sum decomposition. This can be seen most easily by considering the plane which can be direct sum decomposed into two subspaces defined by the horizontal and vertical axes. But the plane can also be direct sum decomposed using subspaces defined by the horizontal axes and any other nonparallel line going through zero.

VI

Periodic Solutions of Weakly Nonlinear Systems

Throughout most of this book we concentrate on techniques which may, in the end, give "numbers" (e.g., explicit error bounds) as well as insight because this is often what the engineer or physicist needs when confronted with a specific problem. The price that must be paid for this requirement is that the techniques may not be applicable in some important cases. To complement these techniques, we introduce in this chapter an approach for weakly nonlinear systems which leads to conclusions of the form "for sufficiently small ϵ there exists. . . ." Though we do not get, e.g., explicit error bounds, much insight is obtained into various aspects of nonlinear phenomena. The weakly nonlinear approach is applicable in many cases where other approaches fail. Though the importance and usefulness of the weakly nonlinear approach is by no means limited to autonomous systems, we shall concentrate on autonomous systems in this chapter.

An interesting application of the weakly nonlinear theory will be given to show how linear methods of stability analysis can fail for nonlinear systems. The example given also shows how the describing function method (harmonic balance of one frequency) can fail to indicate a self-sustained oscillation while the dual-input describing function (harmonic balance of two frequencies) correctly predicts the oscillation.

VI.1 WEAKLY NONLINEAR SYSTEMS

By a weakly nonlinear system, we mean one which is described by the n-vector differential equation

$$\dot{x} = Ax + \epsilon f(t, x, \epsilon), \tag{VI.1-1}$$

where the matrix A is real and constant, $f(x, t, \epsilon)$ is a real n-vector, continuous and periodic in t of period T and possesses continuous second partial derivatives with respect to x and ϵ for all real x and ϵ.† When $\epsilon = 0$, the exact solution of (VI.1-1) may be obtained and one is interested in how the solution changes for $\epsilon \neq 0$. For the sake of simplicity, we shall assume that $A = 0$ in this chapter. The technique becomes more transparent in this case and one can easily understand its extension to the case of nonzero A. Also, it should be pointed out that even in a case with a nonzero A, it is often convenient to use a transformation to reduce it to a case with $A = 0$; our examples will illustrate this.

Note that the case of $A = 0$ immediately violates one of the assumptions of Chapter V, namely that $\dot{x} = Ax$ does not have nontrivial T-periodic solutions. The equation $\dot{x} = 0$ obviously has constant solutions which are periodic of every period. These cases are critical as opposed to the noncritical cases treated in Chapter V.

VI.2 SUCCESSIVE APPROXIMATIONS

A natural approach to finding a periodic solution to

$$\dot{x} = \epsilon f(t, x, \epsilon) \tag{VI.2-1}$$

is to use the method of successive approximations

$$\dot{x}_{i+1} = \epsilon f(t, x_i, \epsilon) \tag{VI.2-2}$$

with $x_0 = a$, an n-vector, which is the solution of (VI.2-1) when $\epsilon = 0$. However, the successive iterates may not be periodic. For example, consider

$$\dot{x}_1 = \epsilon f(t, x_0, \epsilon) = \epsilon f(t, a, \epsilon). \tag{VI.2-3}$$

Unless $f(t, a, \epsilon)$ has zero mean value, i.e.,

---

† For simplicity, we impose conditions stronger than are actually necessary but which suffice for our applications. Also, for notational brevity, we sometimes, as in (VI.1-1), suppress dependence of functions on t.

$$\frac{1}{T} \int_0^T f(t, a, \epsilon) \, dt = 0 \, , \tag{VI.2-4}$$

the integration of (VI.2-3) leads to nonperiodic terms (called *secular terms*).

It is thus clear that the vector a must be judiciously chosen. The approach used is, roughly speaking, the following. We first choose an arbitrary vector a, then use the method of successive approximations casting out the troublesome secular terms (which allows the continuation of the successive approximations with periodic functions). Assuming the successive approximations converge to $x^*(t, a, \epsilon)$, the time of reckoning for casting out secular terms comes when one examines what equation $x^*(t, a, \epsilon)$ satisfies. It turns out that $x^*(t, a, \epsilon)$ does not necessarily satisfy (VI.2-1) but a differential equation with a right-hand side which has a term in addition to the right-hand side of (VI.2-1). This additional term depends on $x^*(t, a, \epsilon)$ and thus on the vector a. The vector a is chosen to make this additional term vanish. This additional term when set to zero comprises what is called the *determining equation* or *bifurcation equation*. The examples in this chapter will give familiar interpretations to the determining equations.

To show the successive approximations converge, we will use the CMT in the Banach space S of n-vectors of real-valued continuous functions on $(-\infty, \infty)$, periodic with period T, with norm

$$\|x\| = \sup_t |x(t)| \tag{VI.2-5}$$

where the single bars on a vector denotes a norm for an n-vector.

The operator P_0 takes the mean value of a function $x \in S$, i.e.,

$$P_0 x = T^{-1} \int_0^T x(t) \, dt \, . \tag{VI.2-6}$$

The following lemma is needed in the analysis to follow.

Lemma. If $g \in S$, then the equation

$$\dot{x} = g(t) \tag{VI.2-7}$$

has a T-periodic solution if and only if $P_0 g = 0$. Furthermore, there is a unique T-periodic solution x^* with $P_0 x^* = 0$. This solution is designated by $x^*(t) = \int g(u) \, du$ and

$$|x^*(t)| \leq \int_0^T |g(u)| \, du \leq T\|g\| \, . \tag{VI.2-8}$$

Proof. Considering each component† of (VI.2-7) separately,

$$\dot{x}_j = g_j(t) \qquad j = 1, 2, \ldots, n, \tag{VI.2-9}$$

it is clear that this equation has a T-periodic solution if and only if $\int_0^T g_j(t)\, dt = 0$. Furthermore, there is only one solution x_j^* with $\int_0^T x_j^*(t)\, dt = 0$. Otherwise, the difference between two different ones would be a constant. Since each has mean value zero, the constant must be zero. To prove (VI.2-8) we need only show that for any g_j there is a ζ_j, which depends on g_j, satisfying $0 \leq \zeta_j \leq T$ such that

$$x_j^*(t) = \int_{\zeta_j}^t g_j(u)\, du. \tag{VI.2-10}$$

For any $\zeta \in [0, T]$, we have

$$x_j^*(t) = x_j^*(\zeta) + \int_\zeta^t g_j(u)\, du. \tag{VI.2-11}$$

Since x_j^* has zero mean value there is a $\zeta_j \in [0, T]$ such that $x_j^*(\zeta_j) = 0$ so that

$$x_j^*(t) = x_j^*(\zeta_j) + \int_{\zeta_j}^t g_j(u)\, du = \int_{\zeta_j}^t g_j(u)\, du. \tag{VI.2-12}$$

Remark. The integral representation $x^*(t) = \int g(u)\, du$ is only symbolic. Note that (IV.2-10) is complicated since the ζ's depend on g. If g_j has the Fourier series,

$$g_j(t) \sim \sum_{k \neq 0} a_k e^{ik\omega t} \qquad T = \frac{2\pi}{\omega}, \tag{VI.2-13}$$

then

$$x_j^*(t) = \sum_{k \neq 0} \frac{a_k}{ik\omega} e^{ik\omega t} \tag{VI.2-14}$$

since this x_j^* satisfies (VI.2-9) and has zero mean value.

The next theorem shows convergence of successive approximation by showing the mapping used in successive approximations is a contraction mapping.

---

† We use subscripts on a vector to denote components of the vector; no confusion with using subscript for an iterate should arise.

The mapping is defined by the right-hand side of (VI.2-1) with the mean value subtracted out [recall the discussion below (VI.2-4)].

For given real numbers b, d, satisfying $b < d$, and a given n-vector a, $|a| \leq b$, the closed set S_0 is defined as follows:

$$S_0 = \{x \in S: \; P_0 x = a, \; \|x\| \leq d\}. \tag{VI.2-15}$$

Theorem VI.2.1

Given a positive number b, there is an $\epsilon_1 > 0$ such that, corresponding to each constant n-vector a, $|a| \leq b$, and to each ϵ, $|\epsilon| \leq \epsilon_1$, there is a unique function

$$x^*(t) = x^*(t, a, \epsilon) \in S_0 \tag{VI.2-16}$$

which has a continuous first derivative with respect to t, and whose derivative satisfies

$$\dot{x}^* = \epsilon f(t, x^*, \epsilon) - \epsilon P_0 f(t, x^*, \epsilon). \tag{VI.2-17}$$

Furthermore, $x^*(t, a, 0) = a$ and $x^*(t, a, \epsilon)$ has continuous first derivatives with respect to a, ϵ.

Proof. For $x \in S_0$ define the mapping F by

$$z = Fx = a + \epsilon \int (I - P_0) f(u, x(u), \epsilon) \, du, \tag{VI.2-18}$$

where the $\int$ operator is as defined in the Lemma. From (VI.2-8),

$$|(Fx)(t)| \leq b + |\epsilon| L_1(\epsilon), \tag{IV.2-19}$$

where $L_1(\epsilon)$ is a number independent of $x \in S_0$ [e.g., $L_1(\epsilon) = 2T \max\limits_{|y| \leq d} \max\limits_{t \in [0,T]}$ $|f(t, y, \epsilon)|$]. Therefore, there exists an $\epsilon'_1 > 0$ such that $\|Fx\| \leq d$ for $0 \leq |\epsilon| \leq \epsilon'_1$ since $b < d$ (i.e., $|\epsilon| L_1$ can be made arbitrarily small so that $b + |\epsilon| L_1 \leq d$). Hence, for $|\epsilon| \leq \epsilon'_1$, F maps S_0 into itself.

If $x, y \in S_0$ and $|\epsilon| \leq \epsilon'_1$,

$$|(Fx)(t) - (Fy)(t)| \leq 2\epsilon L_2 T \|x - y\|, \tag{VI.2-20}$$

where L_2 is a Lipschitz constant for f in $-\infty < t < \infty$, $|x| \leq d$, $0 \leq |\epsilon| \leq \epsilon'_1$. Thus there exists an $\epsilon_1 \leq \epsilon'_1$ such that F is a contraction for $0 \leq |\epsilon| \leq \epsilon_1$ and there is a unique fixed point $x^*(t) = x^*(t, a, \epsilon)$ of F in S_0. It is clear that $x^*(t, a, 0) = a$. The remainder of the proof is given in Hale, pp. 39, 40 (we do not need these arguments anywhere else in the present book).

The preceding theorem shows that the CMT is quite easily applicable for small enough ϵ although the fixed point x^* satisfies (VI.2-17) which differs from the differential equation of interest (VI.2-1) by the term $\epsilon P_0 f(t, x^*(t, a, \epsilon), \epsilon)$. If we can find a vector a which makes this term vanish then we have a solution to (VI.2-1). Hence, we must satisfy

$$P_0 f(t, x^*(t, a, \epsilon), \epsilon) = T^{-1} \int_0^T f(t, x^*(t, a, \epsilon), \epsilon)\, dt = 0 \,. \qquad \text{(VI.2-21)}$$

It is extremely difficult to discuss satisfaction of (VI.2-21), the determining equation, for $\epsilon \neq 0$ since the dependence of x^* on a is complicated. However, when $\epsilon = 0$, $x^*(t, a, 0) = a$ and (VI.2-21) becomes

$$T^{-1} \int_0^T f(t, a, 0)\, dt = 0 \,. \qquad \text{(VI.2-22)}$$

This can usually be solved fairly easily, being the solution of a set of n transcendental equations. We can then use implicit function theorems to show that for ϵ sufficiently small, there is an $a = a(\epsilon)$ satisfying (VI.2-21).

A version of the implicit function theorem that can be used is the following: Let $X_0(a, \epsilon)$ be an n-vector of real-valued functions defined on and possessing continuous partial derivatives on an open set containing $(a_0, 0)$. Assume that $X_0(a_0, 0) = 0$ and that the determinant of $[\partial X_0(a_0, 0)/\partial a]$ is nonzero (i.e., the Jacobian matrix has an inverse). Then for $|\epsilon|$ sufficiently small, there is a function $a(\epsilon)$ with $a(0) = a_0$ and $a(\epsilon)$ continuous at $\epsilon = 0$ such that $X_0(a(\epsilon), \epsilon) = 0$. Hence, the implicit function theorem gives conditions under which the satisfaction of an equation for certain variables also implies satisfaction if the variables are changed by sufficiently small amounts. This is exactly what is needed to show that (VI.2-22) implies (VI.2-21).

Now define

$$X_0(a, \epsilon) = T^{-1} \int_0^T f(t, x^*(t, a, \epsilon), \epsilon)\, dt \,. \qquad \text{(VI.2-23)}$$

Then,

$$\left[\frac{\partial X_0(a, \epsilon)}{\partial a} \right] = T^{-1} \int_0^T \left[\frac{\partial f(t, x^*(t, a, \epsilon), \epsilon)}{\partial x^*} \right]\left[\frac{\partial x^*(t, a, \epsilon)}{\partial a} \right] dt \qquad \text{(VI.2-24)}$$

and since $x^*(t, a, 0) = a$ and $\partial a/\partial a = I$,

$$\left[\frac{\partial X_0(a, 0)}{\partial a} \right] = T^{-1} \int_0^T \left[\frac{\partial f(t, a, 0)}{\partial a} \right] dt. \qquad \text{(VI.2-25)}$$

If there is an n-vector a such that $X_0(a, 0) = 0$ [i.e., (VI.2-22) is satisfied] and $\det [\partial X_0(a, 0)/\partial a] \neq 0$, then, for $|\epsilon|$ sufficiently small, there will be an $a(\epsilon)$

resulting in satisfaction of (VI.2-21) which, in turn, implies existence of a periodic solution $x^*(t, a(\epsilon), \epsilon)$ of (VI.2-1). We shall also call $X_0(a, 0)$ the determining equation.

The above use of the implicit function theorem leads to a result which states that for some quantity "sufficiently small" some conclusion can be drawn (in this case, the existence of a periodic solution). No quantitative information is given as to what is meant by "sufficiently small." It thus becomes clear why we did not bother to keep track of the constants (L_1, L_2, the ϵ's) in the proof of Theorem VI.2.1 as we, in essence, did in the last chapter; in the end, we only settle for a sufficiently small result, anyway.

VI.3 AUTONOMOUS SYSTEMS

In the preceding section, we showed that satisfaction of a determining equation $X_0(a, 0) = 0$ along with the condition det $[\partial X_0(a, 0)/\partial a] \neq 0$ implied the existence of periodic solutions for $|\epsilon|$ sufficiently small. In the analysis, the period T was assumed known and fixed (determined by the periodicity of the differential equation). In the case of autonomous systems the frequency of a self-sustained oscillation will, in general, vary with ϵ. So we must actually consider the determining equation as a function also of ω, $X_0(a(\epsilon), \omega(\epsilon), \epsilon) = 0$. The basic approach is immediately applicable to this case, also. But we need another version of the implicit function theorem for autonomous systems. Previously, we were concerned with the solvability of (VI.2-22) which is actually n equations (since f is an n-vector) depending on n variables (the n components of the vector a) and ϵ. Now we still have n equations but $(n + 1)$ variables in addition to ϵ due to the introduction of ω as a variable. The Jacobian matrix $[\partial X_0(a_0, \omega(0), 0)/\partial(a, \omega)]$ is not square; it is $n \times (n + 1)$.

The implicit function that may be used is: Let $X_0(a, \omega, \epsilon)$ be an n-vector of real-valued continuous functions defined on and possessing continuous partial derivatives on an open set containing $(a_0, \omega_0, 0)$. Assume that

$$X_0(a_0, \omega_0, 0) = 0 \qquad (VI.3-1)$$

and that there is an $n \times n$ nonsingular submatrix of the Jacobian matrix $[\partial X_0(a_0, \omega_0, 0)/\partial(a, \omega)]$. Then, for $|\epsilon|$ sufficiently small, there are functions $a(\epsilon)$ and $\omega(\epsilon)$, both continuous at $\epsilon = 0$, with $a(0) = a_0$ and $\omega(0) = \omega_0$, such that

$$X_0(a(\epsilon), \omega(\epsilon), \epsilon) = 0. \qquad (VI.3-2)$$

(The condition of nonsingularity of the $n \times n$ submatrix means that the Jacobian matrix has rank n). In the examples to be presented, it will be some parameter related to the frequency rather than the frequency itself which will actually be used in the application of the implicit function theorem.

To illustrate the analysis of autonomous systems, we shall investigate the periodic solutions of the following two autonomous differential equations:

$$\ddot{y} + \sigma^2 y + \epsilon y^3 = 0, \tag{VI.3-3}$$

$$\ddot{y} - \epsilon(1 - y^2)\dot{y} + \sigma^2 y = 0, \qquad \epsilon > 0. \tag{VI.3-4}$$

Although both of these equations are second-order scalar equations, the theory of this chapter is clearly not thus limited. However, by studying second-order equations, we can correlate the results with a method which is convenient only for second-order systems (phase-plane plots). Furthermore, these particular equations are quite interesting in themselves.

The first equation is Duffing's equation with no forcing function. The second is called Van der Pol's equation and it has been used to describe vacuum tube oscillators (see, e.g., McLachlan, p. 41). Relating (VI. 3-4) to the linear differential equation $\ddot{y} + d\dot{y} + \sigma^2 y = 0$, where $d\dot{y}$ represents a damping term, we can visualize a physical mechanism for oscillations. When y is small, there is negative damping and oscillations will tend to build up. When y is large, there is positive damping and oscillations will damp out. This suggests that oscillations will tend to some equilibrium amplitude. In fact, since the damping changes from negative to positive when $y^2 = 1$, it would seem that the oscillation should oscillate around a magnitude not far from 1. We will see how the theory of the last section confirms this.

Both equations are of the form

$$\begin{aligned} \dot{x}_1 &= x_2 \\ \dot{x}_2 &= -\sigma^2 x_1 + \epsilon f(x_1, x_2) \end{aligned} \tag{VI.3-5}$$

and for $\epsilon = 0$, the x_1 solution is $c_1 \sin \sigma t + c_2 \cos \sigma t$ with c_1 and c_2 constants. For small ϵ, it seems likely that the solutions will be similar although we shall see that there is a striking difference between the solutions of (VI. 3-3) and (VI. 3-4). We can first see this by looking at phase-plane plots. These are plots of x_2 versus x_1 with x_1 and x_2 related by (VI.3-5). A point in the phase-plane can represent an initial condition and the path through that point shows behavior at future times. From (VI.3-5)

$$\frac{dx_2}{dx_1} = \frac{-\sigma^2 x_1 + \epsilon f(x_1, x_2)}{x_2}. \tag{VI.3-6}$$

If this differential equation can be solved then those solutions (for different initial conditions) will specify the curves in the phase-plane. Even if the equation is not solved, the curves can be sketched in using the slope information of (VI.3-6) (this is called the method of isoclines). Phase-plane techniques are discussed in much more detail in many places (e.g., Hsu and Meyer).

Typical phase-plane plots for Duffing's and Van der Pol's equations are shown in Figure VI.I. It is seen that while there are a family of periodic solutions to Duffing's equation (each corresponding to different initial conditions), Van der Pol's equation has a periodic solution in the vicinity of which all solutions approach. This is consistent with the physical mechanism of oscillation previously given. The periodic solution of Van der Pol's equation is called a limit cycle which is a closed orbit (set of solution points) in the (x_1, x_2) plane such that no other closed orbit can be found arbitrarily close to it (there are also other definitions of limit cycle). Furthermore, it is seen that other paths converge to the periodic solution. Mathematically stated, the periodic solution possesses asymptotic orbital stability which we shall define in Sec. VIII.1.

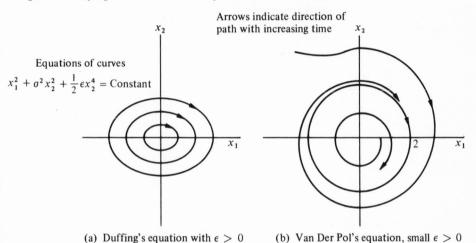

(a) Duffing's equation with $\epsilon > 0$ (b) Van Der Pol's equation, small $\epsilon > 0$

Figure VI.1 Phase-plane plots for Duffing's and Van der Pol's equations

To use the theory of the last section, we introduce the transformation

$$x_1 = z_1 \sin \omega t + z_2 \cos \omega t$$
$$x_2 = \omega(z_1 \cos \omega t - z_2 \sin \omega t) \tag{VI.3-7}$$

with

$$\omega^2 = \sigma^2 + \epsilon\beta. \tag{VI.3-8}$$

We obtain

$$\dot{z}_1 = \frac{\epsilon}{\omega}\left[\beta x_1 + f(x_1, x_2)\right]\cos \omega t$$
$$\dot{z}_2 = -\frac{\epsilon}{\omega}\left[\beta x_1 + f(x_1, x_2)\right]\sin \omega t \tag{VI.3-9}$$

which is in the form of (VI.2-1). From (VI.2-22), the determining equations are, with a_0 the vector (a_1, a_2),

$$F_1(a_1, a_2, \beta) \equiv \frac{1}{2\pi} \int_0^{2\pi/\omega} \big[\beta(a_1 \sin \omega t + a_2 \cos \omega t) + f(a_1 \sin \omega t + a_2 \cos \omega t,$$

$$\omega(a_1 \cos \omega t - a_2 \sin \omega t)) \big] \cos \omega t \, dt = 0 \quad \text{(VI.3-10)}$$

$$F_2(a_1, a_2, \beta) \equiv \frac{1}{2\pi} \int_0^{2\pi/\omega} \big[\beta(a_1 \sin \omega t + a_2 \cos \omega t) + f(a_1 \sin \omega t + a_2 \cos \omega t,$$

$$\omega(a_1 \cos \omega t - a_2 \sin \omega t)) \big] \sin \omega t \, dt = 0. \quad \text{(VI.3-11)}$$

Performing the integrations of (VI.3-10) and (VI.3-11) and setting $\epsilon = 0$ shows that ω may be replaced by σ. It is of interest to see that these equations are consistent with the use of harmonic balance. Putting

$$y = a_1 \sin \omega t + a_2 \cos \omega t \quad \text{(VI.3-12)}$$

into

$$\ddot{y} + \sigma^2 y - \epsilon f(y, \dot{y}) = 0 \quad \text{(VI.3-13)}$$

and using (VI.3-8) yields

$$\epsilon \beta y + \epsilon f(y, \dot{y}) = 0. \quad \text{(VI.3-14)}$$

Balancing the $\cos \omega t$ and $\sin \omega t$ terms on both sides of (VI.3-14) yields (VI.3-10) and (VI.3-11), respectively (for $\epsilon \neq 0$).

For Duffing's equation, it is easily verified that the first determining equation is satisfied for any $(a_1, 0)$ and the second determining equation then reduces to

$$\beta = \tfrac{3}{4} a_1^2. \quad \text{(VI.3-15)}$$

This gives the following approximation to the frequency of the periodic solution

$$\omega^2 \cong \sigma^2 + \tfrac{3}{4}\epsilon a_1^2 \quad \text{(VI.3-16)}$$

[which agrees with (V.3-3) for $f = 0$].

In this case we do not appeal to the implicit function theorem as stated at the beginning of this section. Rather we recognize that $F_1(a_1, 0, \beta) = 0$ for all a_1 and β, and we use an implicit function theorem to show that $F_2(a_1, 0, \beta)$ is satisfied with $a_1(\epsilon)$ and $\beta(\epsilon)$ close to the a_1 and β of (VI.3-15). Hence, in agreement with the phase-plane plots of Figure VI.1, there is a family of periodic solutions. They are approximately describable by amplitudes and frequencies related by (VI.3-16).

The situation with Van der Pol's equation is quite different. In this case, the determining equations (VI.3-10) and (VI.3-11) reduce to

$$F_1(a_1, a_2, \beta) = \beta a_2 + \sigma a_1 [1 - \tfrac{1}{4}(a_1^2 + a_2^2)] = 0$$
$$F_2(a_1, a_2, \beta) = \beta a_1 - \sigma a_2 [1 - \tfrac{1}{4}(a_1^2 + a_2^2)] = 0 \qquad \text{(VI.3-17)}$$

which are satisfied with $a_1^2 + a_2^2 = 4, \beta = 0$. The Jacobian matrix

$$\left[\frac{\partial F_i(a_1, a_2, \beta)}{\partial (a_1, a_2, \beta)}, i = 1, 2 \right] = \begin{bmatrix} \sigma\left(1 - \dfrac{3a_1^2}{4} - \dfrac{a_2^2}{4}\right) & \beta - \dfrac{\sigma a_1 a_2}{2} & a_2 \\[3mm] \beta + \dfrac{\sigma a_1 a_2}{2} & -\sigma\left[1 - \dfrac{a_1^2}{4} - \dfrac{3a_2^2}{4}\right] & a_1 \end{bmatrix}$$

$$\text{(VI.3-18)}$$

is easily seen to have rank 2 with $a_1^2 + a_2^2 = 4, \beta = 0$ (the rank is the size of the largest square submatrix with nonzero determinant). Hence, for ϵ sufficiently small, there is a periodic solution to (VI.3-4) approximately describable by an amplitude of 2 and a frequency of σ. Note that, in contrast to the case for Duffing's equation, both the amplitude and frequency are (approximately) independent of ϵ for small enough ϵ.

VI.4 FAILURE OF DESCRIBING FUNCTIONS; AIZERMAN'S CONJECTURE

A nice application of this chapter's theory is to show how linear methods can fail for nonlinear systems. Consider the unforced feedback loop shown in Figure VI.2. The nonlinear function is assumed to satisfy $n(0) = 0$ and the

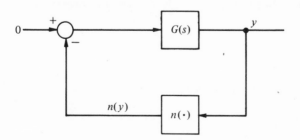

Figure VI.2 System with nonlinear feedback

transfer function $G(s)$ is in the usual Laplace transform notation and is assumed to be rational.

In Chapter VIII we shall present a rigorous approach to stability of nonlinear feedback systems. At this point, we can consider the question of whether the stability can be determined by considering related linear systems. First we must give a definition to stability. For linear feedback systems with rational transfer functions, Nyquist's criterion is a necessary and sufficient condition for the poles of the closed loop transfer function to be in the left half-plane which has strong implications regarding the system's responses. We shall impose here the simple requirement that the nonlinear feedback system's unforced output responses approach zero as $t \to \infty$. If the nonlinear feedback system has this property, we shall (in this section only) call the system stable. We shall discuss stability more comprehensively in Chapter VIII; the stability definition given here is used only for simplicity in this section.

If n is a linear function,† $n(\sigma) = k\sigma$, stability may be determined in a number of ways, e.g., Nyquist's criterion, Hurwitz–Routh criterion. If $G(s)$ has no poles in the right-half plane, Re $s \geq 0$, Nyquist's criterion for the linear feedback system is recalled to be that the locus of $G(i\omega)$ for $\omega \in (-\infty, \infty)$ does not encircle or go through the point $(-k^{-1}, 0)$ in the complex plane. To put Nyquist's criterion into a context more suitable for the present section, say that we wish the linear feedback system to be stable for all linear gains $k \in [0, K)$. A necessary and sufficient condition for stability for all linear gains $k \in [0, K)$ is that the locus of $G(i\omega)$ for $\omega \in (-\infty, \infty)$ does not intersect the real axis

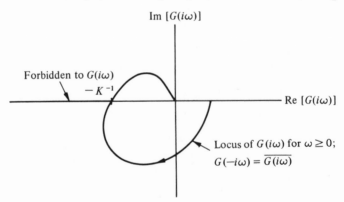

Figure VI.3 Nyquist's criterion for all $k \in [0, K)$

from $-\infty$ to $-K^{-1}$ (see Figure VI.3). Otherwise, the linear system would violate Nyquist's criterion for some $k \in [0, K)$.

With $n(\sigma) = k\sigma$, suppose that the loop is stable for all $k \in [0, K)$. Now

---

† As mentioned in the Preface, we use nonlinear to mean not necessarily linear; it is not defined to be not linear.

suppose that n need not be linear but merely that it is such that existence problems are resolved and that $n(0) = 0$ and

$$0 \leq \frac{n(\sigma)}{\sigma} < K, \qquad \sigma \neq 0. \tag{VI.4-1}$$

The graph of n lies in the sector shown in Figure VI.4, the sector determined from the stability limits for the linear system. The graph of n is the set of ordered pairs $\{(\sigma, n(\sigma)) : \sigma \in (-\infty, \infty)\}$. Aizerman's conjecture is that the system with such a nonlinear n is also stable. Although this conjecture† is quite plausible and it is not that easy to find counterexamples, counterexamples do

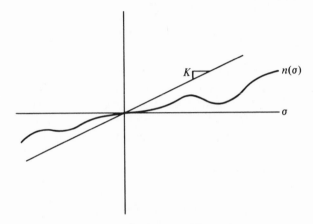

Figure VI.4 The graph of n confined to a sector

exist. This section will produce one. In fact, the counterexample will also show that the stronger condition

$$0 \leq \frac{dn}{d\sigma} < K \tag{VI.4-2}$$

is not sufficient for stability. Condition (VI.4-1) means that the total or "d.c." gain of the nonlinearity is between the stability limits for linear gains. Condition (VI.4-2) refers to the incremental or "a.c." gains.

Another common method of heuristically predicting stability is by the use of describing functions. If no nontrivial self-excited oscillations are predicted with describing functions, the feedback system is (heuristically) concluded to be stable. The describing function method indicates the existence of an oscillation of frequency ω and amplitude E at the input to the nonlinearity if

---

† It may be noted that the linear system may also be stable for negative gains as well as for gains in disjoint sectors.

$G(i\omega)N(E) = -1$ where $N(E)$ is the describing function (see Sec. IV.4). Recalling from a footnote in Sec. IV.4 that $N(E)$ is real, we see that the describing function stability criterion (nonexistence of oscillations) is less stringent than using Aizerman's conjecture since satisfaction of (VI.4-1) implies

$$N(E) = \frac{2T^{-1} \int_0^T n(E \sin \omega t) \sin \omega t \, dt}{E}$$

$$\leq 2T^{-1} \int_0^T K \sin^2 \omega t \, dt = K. \tag{VI.4-3}$$

Similarly,

$$N(E) \geq 2T^{-1} \int_0^T 0 \, dt = 0. \tag{VI.4-4}$$

In order for the system to be concluded to be stable via describing functions for all $k \in [0, K)$, the locus of $G(i\omega)$ for $\omega \in (-\infty, \infty)$ on the complex plane must not pass through the portion of the real axis which is a subset of the segment of the real axis already forbidden to $G(i\omega)$ by Aizerman's conjecture.

Hence, we are presenting a counterexample to three types of stability prediction by linearization.

The specific system to be studied is

$$G(s) = \frac{s^2}{(s^2 + 1)(s^2 + 9) + \epsilon(\alpha s^3 + \beta s^2 + \gamma s + \delta)} \tag{VI.4-5}$$

$$\alpha = 3/16, \qquad \beta = 29/16, \qquad \gamma = 15/16, \qquad \delta = 117/16 \tag{VI.4-6}$$

$$n(\sigma) = \epsilon f(\sigma)$$
$$f(\sigma) = \sigma^3, \qquad |\sigma| \leq \sigma_m,$$
$$= \sigma^3 e^{-(\sigma - \sigma_m)^4}, \qquad \sigma > \sigma_m,$$
$$= \sigma^3 e^{-(\sigma + \sigma_m)^4}, \qquad \sigma < -\sigma_m. \tag{VI.4-7}$$

σ_m is some large number so that f behaves as a cubic in all the calculations below but it prevents the gain $n(\sigma)/\sigma$ from approaching infinity as $|\sigma| \to \infty$ (while it preserves the continuous second derivative). Although a numerical value for σ_m will be needed in Sec. VIII.6, none is needed in the present section.

Letting

$$y(t) = v^{(2)}(t), \qquad v^{(i)} = \frac{d^i v}{dt^i} \tag{VI.4-8}$$

a differential equation for v is

$$v^{(4)} + 10v^{(2)} + 9v + \epsilon(\alpha v^{(3)} + \beta v^{(2)} + \gamma v^{(1)} + \delta v) + \epsilon f(v^{(2)}) = 0.$$
$$\text{(VI.4-9)}$$

Consider the linear differential equation [a linearization of (VI.4-9)]

$$v^{(4)} + 10v^{(2)} + 9v + \epsilon(\alpha v^{(3)} + \beta v^{(2)} + \gamma v^{(1)} + \delta v) + \epsilon k v^{(2)} = 0.$$
$$\text{(VI.4-10)}$$

The Hurwitz–Routh criterion (see, e.g., Kaplan, p. 407) for all of the roots of

$$s^4 + 10s^2 + 9 + \epsilon(\alpha s^3 + \beta s^2 + \gamma s + \delta) + \epsilon k s^2 = 0 \quad \text{(VI.4-11)}$$

to have negative real parts is that the coefficients of all of the powers of s be positive and

$$\Delta_2 \equiv \epsilon\alpha[10 + \epsilon(\beta + k)] - \epsilon\gamma > 0,$$
$$\epsilon\gamma\Delta_2 - (9 + \epsilon\delta)\epsilon^2\alpha^2 > 0. \qquad \text{(VI.4-12)}$$

For any k, $0 \leq k < \infty$, there is a sufficiently small positive ϵ such that the Hurwitz–Routh criterion is satisfied for the parameters given by (VI.4-6). Hence, using any of the three previously mentioned linearizations, the system would be concluded to be stable for sufficiently small ϵ. Note that with $\epsilon = 0$, the roots of (VI.4-11) are imaginary, but a small positive ϵ pushes them into the left half-plane when the linearized systems are considered. With $\epsilon = 0$ the system clearly has periodic solutions; we will show that, for ϵ sufficiently small, the system still has periodic solutions. Hence, we have a counterexample to stability determination by linearization.

The determining equations for the weakly nonlinear theory will turn out to be consistent with those obtained by use of the dual-input describing function method (harmonic balance of two frequencies). Since this method is quite transparent, let us apply it first. Trying an approximate solution of the form $a_1 \sin \omega t + a_2 \cos \omega t + a_3 \sin 3\omega t + a_4 \cos 3\omega t$ with $a_2 = a_3 = 0$ in (VI.4-9) yields after dropping all terms of frequency exceeding 3ω

$$\sin \omega t \left[\omega^4 a_1 - 10\omega^2 a_1 + 9a_1 - \epsilon\beta\omega^2 a_1 + \epsilon\delta a_1 - \frac{\epsilon\omega^6}{4} 3a_1^3 - \epsilon\omega^6 \frac{243a_1 a_4^2}{2} \right]$$

$$+ \cos \omega t \left[-\epsilon\alpha\omega^3 a_1 + \epsilon\gamma\omega a_1 + \epsilon\omega^6 \frac{27a_1^2 a_4}{4} \right]$$

$$+ \sin 3\omega t \left[\epsilon\alpha 27\omega^3 a_4 - \epsilon\gamma 3\omega a_4 + \epsilon\omega^6 \frac{a_1^3}{4} \right]$$

$$+ \cos 3\omega t \left[81\omega^4 a_4 - 90\omega^2 a_4 + 9a_4 - \epsilon\beta 9\omega^2 a_4 + \epsilon\delta a_4 \right.$$

$$\left. - \epsilon\omega^6 \frac{27a_1^2 a_4}{2} - \epsilon\omega^6(\tfrac{3}{4})729a_4^3 \right] = 0. \qquad \text{(VI.4-13)}$$

For simplicity of presentation, we are using hindsight in letting $a_2 = a_3 = 0$ since it turns out these can be set to zero.† Letting

$$\omega^2 = 1 + \epsilon b \qquad (VI.4\text{-}14)$$

we observe that (VI.4-13) contains terms of frequency ω and 3ω which are multiplied by different powers of ϵ. The usual dual-input describing function method would balance the coefficients of the sin ωt, cos ωt, sin $3\omega t$, and cos $3\omega t$ terms, ignoring the higher harmonics. Since we are interested in the case when $\epsilon \to 0$, we shall also ignore all terms multipled by ϵ^p, $p \geq 2$. We then get the following conditions for balance of the sin ωt, cos ωt, sin $3\omega t$ and cos $3\omega t$ terms, respectively:

$$a_1 - 10a_1 + 9a_1 + \epsilon\left(-8ba_1 - \beta a_1 + \delta a_1 - \tfrac{3}{4}a_1^3 - \frac{243}{2}a_1 a_4^2\right) = 0 \tag{VI.4-15}$$

$$\epsilon\left(-\alpha a_1 + \gamma a_1 + \frac{27a_1^2 a_4}{4}\right) = 0 \tag{VI.4-16}$$

$$\epsilon\left(27\alpha a_4 - 3\gamma a_4 + \frac{a_1^3}{4}\right) = 0 \tag{VI.4-17}$$

$$81a_4 - 90a_4 + 9a_4 + \epsilon\left(72ba_4 - 9\beta a_4 + \delta a_4 - \frac{27}{2}a_1^2 a_4 - \tfrac{3}{4}729a_4^3\right) = 0. \tag{VI.4-18}$$

With the parameters of (VI.4-6), the last four equations are satisfied with $a_1 = -1, a_4 = \tfrac{1}{9}, b = 13/32$. Hence, the dual-input describing function method suggests that, for small ϵ, (VI.4-9) has a periodic solution $v(t, \epsilon)$ such that

$$\lim_{\epsilon \to 0} v(t, \epsilon) = -\sin t + \tfrac{1}{9}\cos 3t. \tag{VI.4-19}$$

The period is approximately given by

$$T = \frac{2\pi}{\omega} \cong \frac{2\pi}{\sqrt{1 + \epsilon b}} \cong 2\pi\left(1 - \frac{13}{64}\epsilon\right). \tag{VI.4-20}$$

Note that in this case the use of the dual-input describing function (balancing terms of frequencies ω and 3ω) contradicts the use of the single-input describing function (balancing terms of frequency ω) which does not predict

---

† Actually, we need never consider a nonzero a_2 since $a_1 \sin \omega t + a_3 \sin 3\omega t + a_4 \cos 3\omega t$ is as general as is required except for an unimportant phase shift. Some of the manipulations of the previous section could have been slightly simplified using a similar observation.

any self-oscillations. The following application of the weakly nonlinear theory will confirm the dual-input describing function result.

We first put the system's equations into a form suitable for application of the weakly nonlinear theory. Letting

$$w_1 = v, \qquad \dot{w}_1 = w_2, \qquad \dot{w}_2 = w_3, \qquad \dot{w}_3 = w_4, \qquad \text{(VI.4-21)}$$

we obtain the following differential equation for the 4-vector w,

$$\dot{w} = \begin{bmatrix} 0 & 1 & 0 & 0 \\ 0 & 0 & 1 & 0 \\ 0 & 0 & 0 & 1 \\ -9 & 0 & -10 & 0 \end{bmatrix} w - \epsilon \begin{bmatrix} 0 & 0 & 0 & 0 \\ 0 & 0 & 0 & 0 \\ 0 & 0 & 0 & 0 \\ \delta & \gamma & \beta & \alpha \end{bmatrix} w - \epsilon \begin{bmatrix} 0 \\ 0 \\ 0 \\ f(w_3) \end{bmatrix}.$$

$$\text{(VI.4-22)}$$

We now change variables with the transformation† $w = Px$ with

$$P = \begin{bmatrix} 1 & 0 & 1 & 0 \\ 0 & 1 & 0 & 3 \\ -1 & 0 & -9 & 0 \\ 0 & -1 & 0 & -27 \end{bmatrix}, \qquad P^{-1} = \tfrac{1}{8} \begin{bmatrix} 9 & 0 & 1 & 0 \\ 0 & 9 & 0 & 1 \\ -1 & 0 & -1 & 0 \\ 0 & -\tfrac{1}{3} & 0 & -\tfrac{1}{3} \end{bmatrix}.$$

$$\text{(VI.4-23)}$$

The new differential equation in x is [using the oddness of f, $f(-\sigma) = -f(\sigma)$]

$$\dot{x} = \begin{bmatrix} 0 & 1 & 0 & 0 \\ -1 & 0 & 0 & 0 \\ 0 & 0 & 0 & 3 \\ 0 & 0 & -3 & 0 \end{bmatrix} x - \frac{\epsilon}{8} \begin{bmatrix} 0 & 0 & 0 & 0 \\ \delta - \beta & \gamma - \alpha & \delta - 9\beta & 3(\gamma - 9\alpha) \\ 0 & 0 & 0 & 0 \\ \dfrac{\beta - \delta}{3} & \dfrac{\alpha - \gamma}{3} & \dfrac{9\beta - \delta}{3} & 9\alpha - \gamma \end{bmatrix} x$$

$$+ \frac{\epsilon}{8} \begin{bmatrix} 0 \\ 1 \\ 0 \\ -\tfrac{1}{3} \end{bmatrix} f(x_1 + 9x_3). \qquad \text{(VI.4-24)}$$

This shows the system to represent two harmonic oscillators coupled by terms

† This type of tranformation is related to a similarity transformation of a matrix with distinct eigenvalues to a diagonal matrix with the eigenvalues on the diagonal. With complex eigenvalues, one may wish to transform instead to a real matrix (which is what the above does). Usually the P matrix is selected with ones in the first row but this is somewhat arbitrary.

of order ϵ which is also apparent from inspection of the transfer function given by (VI.4-5). To put it into final form for application of the theory we use the following transformation from the vector x to the vector z

$$x_1 = z_1 \sin \omega t + z_2 \cos \omega t$$
$$x_2 = \omega(z_1 \cos \omega t - z_2 \sin \omega t)$$
$$x_3 = z_3 \sin 3\omega t + z_4 \cos 3\omega t \qquad \text{(VI.4-25)}$$
$$x_4 = \omega(z_3 \cos 3\omega t - z_4 \sin 3\omega t).$$

Having gone through a couple of transformations, it is useful at this point to trace back from the present state variables to the solution of the original differential equation (VI.4-9):

$$v = w_1 = x_1 + x_3$$
$$\dot{v} = w_2 = x_2 + 3x_4. \qquad \text{(VI.4-26)}$$

Now letting $\omega^2 = \sigma^2 + \epsilon b$, we obtain

$$\dot{z}_1 = \frac{\epsilon}{\omega} \{bx_1 - \tfrac{1}{8}[\delta - \beta, \gamma - \alpha, \delta - 9\beta, 3(\gamma - 9\alpha)]x + \tfrac{1}{8}f(x_1 + 9x_3)\} \cos \omega t$$

$$\dot{z}_2 = -\frac{\epsilon}{\omega} \{bx_1 - \tfrac{1}{8}[\delta - \beta, \gamma - \alpha, \delta - 9\beta, 3(\gamma - 9\alpha)]x + \tfrac{1}{8}f(x_1 + 9x_3)\} \sin \omega t$$

$$\dot{z}_3 = \frac{\epsilon}{\omega} \left\{ 3bx_3 - \frac{1}{8} \left[\frac{\beta - \delta}{3}, \frac{\alpha - \gamma}{3}, \frac{9\beta - \delta}{3}, 9\alpha - \gamma \right]x \right.$$
$$\left. - \frac{1}{24}f(x_1 + 9x_3) \right\} \cos 3\omega t \qquad \text{(VI.4-27)}$$

$$\dot{z}_4 = -\frac{\epsilon}{\omega} \left\{ 3bx_3 - \frac{1}{8} \left[\frac{\beta - \delta}{3}, \frac{\alpha - \gamma}{3}, \frac{9\beta - \delta}{3}, 9\alpha - \gamma \right]x \right.$$
$$\left. - \frac{1}{24}f(x_1 + 9x_3) \right\} \sin 3\omega t.$$

[It is convenient to leave the right hand sides of (VI.4.27) in terms of x.] Letting $\sigma = 1$, $z_1 = a_1$, $z_2 = a_2$, $z_3 = a_3$, $z_4 = a_4$, the determining equations are obtained by integrating $1/\epsilon$ times the right-hand sides of (VI.4-27) over a period $= 2\pi/\omega \simeq 2\pi$ and setting $\epsilon = 0$. This gives four equations which we denote by $F_i(a_1, a_2, a_3, a_4, b) = 0$, $i = 1, 2, 3, 4$. For $a_2 = a_3 = 0$, the determining equations are easily recognized to be the dual-input describing function equations (VI.4-15) to (VI.4-18). Hence, the determining equations are satisfied by $a_1 = -1$, $a_4 = \tfrac{1}{9}$, $b = 13/32$. Furthermore, the Jacobian matrix

$$\left[\frac{\partial F_i(a_1, a_2, a_3, a_4, b)}{\partial(a_1, a_2, a_3, a_4, b)} \right]_{a_1 = -1, \, a_2 = a_3 = 0, \, a_4 = 1/9, \, b = 13/32} \qquad \text{(VI.4-28)}$$

has rank 4 as can be verified in a straightforward manner. Thus, the system does have (nontrivial) self-oscillations satisfying (VI.4-19) and (VI.4-20) for sufficiently small ϵ. This constitutes a counterexample to the three types of linearizations discussed.

Although the counterexample convincingly shows that stability prediction by linearization cannot be generally reliable, it may be that the counterexamples are exceptional cases and that linearization methods might still be fairly reliable guides to stability. However, until the scope of their applicability is delineated (an important open problem), such linearization methods of predicting stability must be approached with great caution, if not avoided altogether. We shall derive rigorous stability conditions for nonlinearities satisfying $0 \leq n(\sigma)/\sigma \leq \beta$ for $\sigma \neq 0$ in Chapter VIII. Aizerman's conjecture will be reconsidered then; it will be an upper bound on the range of applicability of the stability conditions.

VI.5 NOTES

This chapter is based on material in Hale (particularly Chapters 6 and 7) where the reader is referred to for generalizations and for more information on weakly nonlinear systems (including stability). There are other procedures (e.g., power series expansions in ϵ) which lead to determining equations (see, e.g., Stern, Chapter 11).

There are many appropriate references on implicit function theorems, e.g., Fleming, Chapter 4; also see Taylor, Chapter VIII. Implicit function theorems are discussed in the next chapter but in a more general setting than needed for the present chapter.

Phase-plane plots for Duffing's equation are discussed on pp. 274–277 of Hochstadt and those for Van der Pol's equation on pp. 148–151 of McLachlan. Van der Pol's equation is considered using both describing functions and phase-plane plots on pp. 182–188 of Hsu and Meyer.

Aizerman's conjecture is discussed in Chapter V, Sec. 5 of Aizerman and Gantmacher and in Hsu and Meyer. The example in Sec. VI.4 is a slight modification of the one given in Willems [1].

VII

Implicit Functions and Accuracy

of Linearization

We did not dwell too much on the use of implicit function theorems in the last chapter, although they actually played key roles. The present chapter puts implicit function theorems in the spotlight. We shall find them useful in analyzing linearizations and perturbation techniques, e.g., the method of linearization about a solution to a nonlinear differential equation briefly discussed in the Introduction.

An implicit function theorem answers the following question: if $P(x_0, y_0) = 0$, is there a function $F(x) = y$ such that $P(x, F(x)) = P(x, y) = 0$ in a neighborhood of x_0? This formulation can include problems involving fixed points if we let x be an operator and write $P(x, y) = y - x(y)$; if (x_0, y_0) is a zero of $P(x, y)$ then y_0 is a fixed point of x_0.

Our objective is to discuss linearizations and fixed point problems from the point of view of implicit functions. Before doing so, it is first necessary to discuss product spaces and partial derivatives in Banach spaces.

VII.1 PRODUCT SPACES AND PARTIAL DERIVATIVES

Let X and Y be two normed linear spaces. In Sec. I.1 we defined the product set $X \times Y$ as the set of all ordered pairs (x, y) with $x \in X$ and $y \in Y$. If addition of elements of $X \times Y$ and multiplication by a scalar are defined in the natural way,

$$(x_1, y_1) + (x_2, y_2) = (x_1 + y_1, x_2 + y_2) \qquad \text{(VII.1-1)}$$

$$\alpha(x, y) = (\alpha x, \alpha y) \qquad \text{(VII.1-2)}$$

then $X \times Y$ becomes a linear space. We can define a norm on $X \times Y$ by any of the following:

$$\|(x, y)\| = (\|x\|^p + \|y\|^p)^{1/p}, \text{ any } p \geq 1, \qquad \text{(VII.1-3)}$$

$$\|(x, y)\| = \max(\|x\|, \|y\|). \qquad \text{(VII.1-4)}$$

Note that we are not notationally distinguishing between norms in different spaces, the usage being clear ($\|x\|$ for $x \in X$, $\|y\|$ for $y \in Y$, etc). It can be shown that with any of the above norms, $X \times Y$ is a Banach space if X and Y are.

We can identify pairs of the form $(x, 0)$ (where 0 is here the zero element of Y) with elements of X and then X may be considered to be a subspace of $X \times Y$.[†] Y may be analogously considered to be a subspace of $X \times Y$. Every element $(x, y) \in X \times Y$ can then be expressed uniquely as

$$(x, y) = (x, 0) + (0, y) \qquad \text{(VII.1-5)}$$

where the 0's represent the zero elements of X and Y, respectively. Every bounded linear operator $U \in L(X \times Y, Z)$ induces a pair (U_X, U_Y) of bounded linear operators ($U_X \in L(X, Z)$, $U_Y \in L(Y, Z)$):

$$U_X(x) = U((x, 0)), \qquad U_Y(y) = U((0, y)). \qquad \text{(VII.1-6)}$$

Conversely, if $U_X \in L(X, Z)$, $U_Y \in L(Y, Z)$, then $U \in L(X \times Y, Z)$ with

$$U((x, y)) = U_X(x) + U_Y(y). \qquad \text{(VII.1-7)}$$

Now let P be an operator mapping an open subset Ω of a Banach space $X \times Y$ into a Banach space Z. We denote $P((x, y))$ by $P(x, y)$. Let $(x, y_0) \in \Omega$ for some x. The operator defined by

† More precisely, there is a one-one linear mapping of X onto $\{(x, 0) : x \in X\}$.

$$P^{(y_0)}(x) = P(x, y_0) \qquad \text{(VII.1-8)}$$

maps $\Omega^{(y_0)} = \{x : (x, y_0) \in \Omega\}$ into Z. The operator $P^{(x_0)}$ is similarly defined. If $x_0 \in \Omega^{(y_0)}$ and if for all $x \in X$,

$$\lim_{\mu \to 0} \frac{P^{(y_0)}(x_0 + \mu x) - P^{(y_0)}(x_0)}{\mu} = P^{(y_0)\prime}(x_0)x, \qquad \text{(VII.1-9)}$$

where $P^{(y_0)\prime}(x_0) \in L(X, Z)$ then we write $P'_x(x_0, y_0) = P^{(y_0)\prime}(x_0)$ and call $P'_x(x_0, y_0)$ a partial derivative. We similarly define $P'_y(x_0, y_0)$. If P has a derivative $P'(x_0, y_0)$ at $(x_0, y_0) \in \Omega$, then

$$P'(x_0, y_0)(x, y) = P'_x(x_0, y_0)x + P'_y(x_0, y_0)y. \qquad \text{(VII.1-10)}$$

We now have a generalization of the notion of partial derivative given in elementary calculus. It is important to keep in mind that the partial derivatives are bounded linear operators.

VII.2 AN IMPLICIT FUNCTION THEOREM

Let X, Y, Z and $X \times Y$ be Banach spaces. Let P be a continuous map of an open subset $\Omega \subset X \times Y$ into Z. Suppose it is known for some $(x_0, y_0) \in \Omega$ that $P(x_0, y_0) = 0$ (the zero element of Z). The implicit function theorem is concerned with the following question: Is there a function $F(x)$ such that $P(x, F(x)) = 0$ in a neighborhood of x_0? The standard implicit function theorem for Banach spaces gives sufficient conditions for a "yes" answer. The theorem, roughly speaking, says that, assuming $P'_y(x_0, y_0)$ has a bounded linear inverse, then if δ is sufficiently small and $\|x - x_0\| < \delta$, then $P(x, y) = P(x, F(x)) = 0$ and $\|y - y_0\| \le \epsilon$. The theorem does not show how to find explicit δ and ϵ, but only guarantees their existence if δ is sufficiently small. The reader may look back at the implicit function theorem stated in Sec. VI.2 to see that it is a special case.

We shall find it useful to supplement the hypotheses of the standard implicit function theorem so that satisfying the theorem's conditions directly provides one with explicit ϵ and δ (not just "sufficiently small").

Theorem VII.2.1

X, Y, Z, and $X \times Y$ are Banach spaces. Let P be a continuous operator mapping an open subset $\Omega \subset X \times Y$ into Z and $(x_0, y_0) \in \Omega$. Assume:

(1) $P(x_0, y_0) = 0$.

(2) P'_y exists and is continuous in Ω.

(3) The operator $P'_y(x_0, y_0) \in L(Y, Z)$ has a bounded linear inverse $\Gamma = [P'_y(x_0, y_0)]^{-1} \in L(Z, Y)$; $\|\Gamma\| \leq k_1$.

(4) $S = \{(x, y) : \|x - x_0\| < \delta, \|y - y_0\| \leq \epsilon\} \subset \Omega$.

(5) There is a real-valued function $g_1(u, v)$ defined for $u \in [0, \delta]$, $v \in [0, \epsilon]$ and nondecreasing in each argument with the other fixed such that $(x, y) \in S \Rightarrow \|P'_y(x, y) - P'_y(x_0, y_0)\| \leq g_1(\|x - x_0\|, \|y - y_0\|)$.

(6) There is a nondecreasing function g_2 defined on $[0, \delta]$ such that

$$(x, y) \in S \Rightarrow \|P(x, y_0)\| \leq g_2(\|x - x_0\|).$$

(7) $\left\{ \begin{matrix} k_1 g_1(\delta, \epsilon) \leq \gamma < 1 \\ k_1 g_2(\delta) \leq \epsilon(1 - \gamma) \end{matrix} \right\}.$

Then an operator F exists defined for $\|x - x_0\| < \delta$ and mapping $\{x \in X : \|x - x_0\| < \delta\}$ into $\{y \in Y : \|y - y_0\| \leq \epsilon\}$ and possessing the following properties:

(a) $P(x, F(x)) = 0$ $(\|x - x_0\| < \delta)$

(b) $F(x_0) = y_0$

(c) $y = F(x)$ depends continuously on x for $\|x - x_0\| < \delta$.

The operator F is uniquely defined by the properties (a)–(c) in the sense that if F_1 is another operator with the same properties then for all x satisfying $\|x - x_0\| < \delta$ such that $\|F_1(x) - y_0\| \leq \epsilon$, it follows that $F_1(x) = F(x)$.

Proof. Let $(x, y) \in S$. Define $Q^{(x)}(y) = y - \Gamma P(x, y)$. Then,

$$Q^{(x)'}(y) = I - \Gamma P'_y(x, y) = -\Gamma[P'_y(x, y) - P'_y(x_0, y_0)].$$

Hence,

$$\|Q^{(x)'}(y)\| \leq \|\Gamma\| \cdot \|P'_y(x, y) - P'_y(x_0, y_0)\| \leq k_1 g_1(\|x - x_0\|, \|y - y_0\|)$$
$$\leq k_1 g_1(\delta, \epsilon) \leq \gamma < 1.$$

With $(x, y) \in S$ and using the triangle inequality and (II.2-3), we also have

$$\|Q^{(x)}(y) - y_0\| \leq \|\Gamma P(x, y_0)\| + \|Q^{(x)}(y) - Q^{(x)}(y_0)\|$$
$$\leq k_1 g_2(\|x - x_0\|) + \gamma\|y - y_0\|$$
$$\leq k_1 g_2(\delta) + \gamma\epsilon \leq \epsilon(1 - \gamma) + \gamma\epsilon = \epsilon.$$

Thus, if $\|x - x_0\| < \delta$, $Q^{(x)}(y)$ maps the ball defined by $\|y - y_0\| \leq \epsilon$ into itself and is a contraction on that ball. Hence, there exists in this ball a unique fixed point $y^* = F(x)$ of the operator $Q^{(x)}(y)$, i.e., $y^* = y^* - \Gamma P(x, y^*)$ or $P(x, y^*) = 0$. Condition (c) follows from Theorem II.1.4. The uniqueness statement follows from the uniqueness of the fixed point in the ball $\|y - y_0\| \leq \epsilon$.

Remarks. This theorem should be compared with the more standard implicit function theorem (Kantorovich and Akilov, pp. 686, 687). Beyond any difficulty involved in applying the standard implicit function theorem (finding the partial derivative, inverse, etc.) applying the above theorem is often rather straightforward [bounding functions by nondecreasing functions and satisfying condition (7)]. The additional work gives explicit ϵ and δ.

Example VII.2.1

Let $P(x, y) = x^2 + y^2 - 1$ (x and y are real variables). $P(x_0, y_0) = 0$ with $x_0 = 0$, $y_0 = 1$, Application of Theorem VII.2.1 is as follows:

$$P'_y(x, y) = 2y$$
$$k_1 = |(2y_0)^{-1}| = \tfrac{1}{2}$$
$$|P'_y(x, y) - P'_y(x_0, y_0)| = 2|y - y_0| \equiv g_1(|x - x_0|, |y - y_0|)$$
$$|P(x, y_0)| = |x^2 + y_0^2 - 1| = |x - x_0|^2$$
$$\equiv g_2(|x - x_0|), \qquad (x_0 = 0).$$

Condition (7) of the theorem is then

$$\left\{ \begin{matrix} \epsilon \leq \gamma < 1 \\ \tfrac{1}{2}\delta^2 \leq \epsilon(1 - \gamma) \end{matrix} \right\}.$$

For $\delta = 1/\sqrt{2}$, the above is satified with $\epsilon = \gamma = \tfrac{1}{2}$. Hence, the theorem states that if $|x - x_0| \leq 1/\sqrt{2}$, then $|y - y_0| \leq \tfrac{1}{2}$. Actually, $|y - y_0| \leq 0.293$. For $|x - x_0|$ smaller, the results become more accurate. In less trivial examples, the exact information is, of course, usually unavailable.

More valuable information on the operator F may be obtained if P'_x exists and is continuous in Ω. Then F has a Fréchet derivative given by

$$F'(x_0) = -[P'_y(x_0, y_0)]^{-1}P'_x(x_0, y_0) = -\Gamma P'_x(x_0, y_0) \qquad \text{(VII.2-1)}$$

(see Appendix VII.B).[†] Equation (VII.2-1) suggests a natural first approximation for small changes in y due to small changes in x,

$$y \cong y_0 + F'(x_0)(x - x_0). \qquad \text{(VII.2-2)}$$

Relation (VII.2-2) is actually the basis for many commonly used approximations, linearizations, perturbation methods, etc. The next section discusses one

---

† This derivative is easily motivated in elementary calculus. Consider $p(x, y) = 0$. Then $\partial p/\partial x + (\partial p/\partial y) \, dy/dx = 0$ or $dy/dx = -(\partial p/\partial y)^{-1}(\partial p/\partial x)$ *assuming* there is such a differentiable function y.

such linearization in detail. The justification for such linearizations is usually based on changes being "sufficiently small." However, it is usually difficult to quantitatively specify what is meant by "sufficiently small for the validity of linearization." The material in this chapter provides an approach to this problem. In particular, a bound on $\|y - y_0 - F'(x_0)(x - x_0)\|$ is desired. It is shown in Appendix VII.A that

$$
\|y - y_0 - F'(x_0)(x - x_0)\| \leq \|\Gamma\| \Big[\sup_{\substack{\|x - x_0\| < \delta_1 \\ \|y - y_0\| \leq \epsilon_1}} \|P'_x(x, y) - P'_x(x_0, y_0)\|
$$

$$
\cdot \|x - x_0\| + \sup_{\substack{\|x - x_0\| < \delta_2 \\ \|y - y_0\| \leq \epsilon_2}} \|P'_y(x, y) - P'_y(x_0, y_0)\| \cdot \|y - y_0\| \Big],
$$

$$(VII.2-3)$$

where $(\delta_1, \epsilon_1, \delta_2, \epsilon_2)$ is either $(\delta, \epsilon, 0, \epsilon)$ or $(\delta, 0, \delta, \epsilon)$ whichever gives the lower bound and δ and ϵ are determined from Theorem VII.2.1.

Remark. The bound on $\|\Gamma\|$ and the g_1 function from Theorem VII.2.1 may be used for (VII.2-3).

The next section shows that a commonly used method of linearization around a nominal solution to a nonlinear differential equation (as mentioned in the Introduction) is equivalent to using (VII.2-1) and (VII.2-2) and a bound on the error due to linearization is obtained for an example.

Before we go on to the next section, it is instructive to see how the rather abstract results of this section look when specialized to a much simpler problem. Consider the equation

$$
y = f(x) \tag{VII.2-4}
$$

where f is a real-valued continuously differentiable function of the real variable x. If $y_0 = f(x_0)$, then a first approximation for changes in y due to changes in x is

$$
y \cong y_0 + f'(x_0)(x - x_0). \tag{VII.2-5}
$$

We can put this into our more general setting by letting

$$
P(x, y) = y - f(x) = 0. \tag{VII.2-6}
$$

Note that in (VII.2-6), y is expressed as an *explicit* function of x while our more general discussion did not assume this. Indeed, the implicit function theorem is used to prove merely the existence of an *implicit* function $F(x) = y$. The knowledge of the explicit function $F(x)$ considerably simplifies matters.

Exercise VII.2.1

Show how to apply Theorem VII.2.1 and (VII.2-3) to the simple problem discussed above. In particular, show that

$$|y - y_0 - f'(x_0)(x - x_0)| \leq \max_{|x-x_0|\leq\delta} |f'(x) - f'(x_0)| \cdot |x - x_0|.$$

$$\text{(VII.2-7)}$$

Interpret (VII.2-7). Must Theorem VII.2.1 be used to obtain VII.2-7? Can (VII.2-7) be generalized?

This simple problem is not without practical interest. For example, consider the following network problem. Let $G(s, p)$ be a transfer function with $s = \sigma + i\omega$ and p a real parameter. For simplicity, consider s fixed. Say $G(s, p)$ is known for some nominal p_0. How does $G(s, p)$ vary with small changes in p? Typicany one does the following:

$$G(s, p) \cong G(s, p_0) + \frac{\partial G(s, p_0)}{\partial p}(p - p_0) .$$

$$\text{(VII.2-8)}$$

The user of (VII.2-8) is fairly confident that it is accurate for $(p - p_0)$ "sufficiently small" but quantitative determination of what is meant by sufficiently small is usually left open. The above approach is applicable to determination of what is meant by sufficiently small; (VII.2-7) holds for complex-valued functions.

In the previous paragraph, we assumed s was some fixed complex number and we consequently get a bound on the accuracy of linearization for that s. We could then consider the bound as a function of s. Alternatively, we might bound

$$\max_{s \in S} \left| G(s, p) - G(s, p_0) - \frac{\partial G(s, p_0)}{\partial p}(p - p_0) \right|,$$

$$\text{(VII.2-9)}$$

where S is some set of complex numbers if we recognize that $\max_{s \in S} | \cdot |$ can actually define a norm.

One further observation is worthwhile. If f has a bounded second derivative, the right-hand side of (VII.2-7) can be bounded as follows:

$$\max_{|x-x_0|\leq\delta} |f'(x) - f'(x_0)| \cdot |x - x_0| \leq \sup_{|x-x_0|\leq\delta} |f''(x)| \cdot |x - x_0|^2, \quad \text{(VII.2-10)}$$

using the mean value formula (I.6-2) on the first derivative. Second derivatives can also be generalized to Banach spaces (Kantorovich and Akilov, pp. 668–676) and are useful in more abstract problems.

VII.3 ACCURACY OF LINEARIZATION; THE EQUATION OF FIRST VARIATION

Consider the vector differential equation

$$\dot{y}(t) = f(y(t), x), \ y(0) = 0, \ t \in [0, t_1] = T, \qquad \text{(VII.3-1)}$$

where x is a vector parameter. Suppose for some x_0 that a solution y_0 is obtained for the differential equation. What if x changes from x_0? How does the solution y vary (assuming it still exists)? If $\delta x = x - x_0$ is "small," the approximate change in y, δy, is often obtained from the solution of the equation of first variation,

$$\delta \dot{y}(t) = f_y(y_0(t), x_0)\delta y(t) + f_x(y_0(t), x_0)\delta x, \qquad t \in T, \qquad \delta y(0) = 0,$$
$$\text{(VII.3-2)}$$

where f_y and f_x are Jacobian matrices of partial derivatives (i.e., the ij'th component of $f_y(y_0(t), x_0)$ is $\partial f_i(y_0(t), x_0)/\partial y_j$ where the i and j subscripts refer to components of vectors). The derivatives are assumed to exist and to be continuous. Note that (VII.3-2) is a linear (time-varying, in general) differential equation and can be solved using transition matrices.

The use of (VII.3-2) is a special case of (VII.2-1) and (VII.2-2). To show this, convenient Banach spaces are first defined. If the parameter x is a real-valued q-vector, let X be a q-dimensional real space with a convenient norm. If y is an r-vector of real-valued time functions on $T = [0, t_1]$, let Y be a space of r-vectors of real-valued continuously differentiable functions of time on T with initial condition $y(0) = 0$ (for a linear space) and with a convenient norm (an example will be given in this section). Z is a space of r-vectors of real-valued continuous functions of time on T with a convenient norm.

Now consider the operator P mapping a subset of $X \times Y$ into Z:

$$z = P(x, y) \qquad \text{(VII.3-3)}$$

$$z(t) = \dot{y}(t) - f(y(t), x), \qquad t \in T. \qquad \text{(VII.3-4)}$$

When $z = 0$, we clearly have a solution to the differential equation (VII.3-1). It may be shown that for many spaces and under mild conditions on f that the partial derivative operator $z = P'_y(x_0, y_0)y$ is defined by

$$z(t) = \dot{y}(t) - f_y(y_0(t), x_0)y(t), \qquad t \in T. \tag{VII.3-5}$$

Note that we are using z essentially as a dummy variable. In order to use (VII.2-1) and (VII.2-2) the inverse of $P_y'(x_0, y_0)$ is needed. This is obtained from the solution of (VII.3-5) for $y(t)$ in terms of $z(t)$. The solution of (VII.3-5) is [with $y(0) = 0$]

$$y(t) = \int_0^t \Phi(t, s)z(s)\, ds, \qquad t \in T, \tag{VII.3-6}$$

where $\Phi(t, s)$ is the transition matrix for

$$\dot{y}(t) = f_y(y_0(t), x_0)y(t). \tag{VII.3-7}$$

To use (VII.2-1) and (VII.2-2), the partial derivative operator $z = P_x'(x_0, y_0)x$ must still be found. This is given by (under mild conditions)

$$z(t) = -f_x(y_0(t), x_0)x, \qquad t \in T. \tag{VII.3-8}$$

Putting (VII.2-1), (VII.2-2), (VII.3-6), and (VII.3-8) together, letting $y - y_0 = \delta y$, $x - x_0 = \delta x$ and using equality instead of approximate equality in (VII.2-2) yields

$$\delta y(t) = \int_0^t \Phi(t, s)f_x(y_0(s), x_0)\delta x\, ds, \qquad t \in T, \tag{VII.3-9}$$

or

$$\delta \dot{y}(t) = f_y(y_0(t), x_0)\delta y(t) + f_x(y_0(t), x_0)\delta x, \qquad \delta y(0) = 0, \qquad t \in T, \tag{VII.3-10}$$

the equation of first variation.

In deriving (VII.3-10), it was mentioned that the approximate equality of (VII.2-2) was replaced by equality. We wish to estimate the accuracy of this approximation. In particular, a bound on $\|y - y_0 - F'(x_0)(x - x_0)\|$ is desired. Such information is obtainable from (VII.2-3). But first the appropriate δ and ϵ must be found using Theorem VII.2.1.

To fix ideas, we detail here a simple illustrative example, one which is solvable exactly so that an assessment of the method in this case is easily carried out. Consider the scalar differential equation

$$\dot{m}(t) + m(t) + xm^3(t) = 0, \qquad m(0) = 1, t \in [0, 1] = T, \tag{VII.3-11}$$

where x is a parameter. Let $y(t) = m(t) - 1$ so that $y(0) = 0$ (for a linear space). Then

$$\dot{y}(t) + y(t) + 1 + x(y(t) + 1)^3 = 0, \qquad y(0) = 0, \qquad t \in T.$$

$$(VII.3-12)$$

If $x_0 = 0$, then

$$y_0(t) = e^{-t} - 1, \qquad t \in T, \qquad\qquad (VII.3-13)$$

and the equation of first variation (with $\delta x = x$ since $x_0 = 0$) is

$$\delta \dot{y}(t) = -\delta y(t) - e^{-3t} x, \qquad \delta y(0) = 0, \qquad t \in T. \qquad (VII.3-14)$$

To use Theorem VII.2.1, let $z = P(x, y)$ be defined by

$$z(t) = \dot{y}(t) + y(t) + 1 + x(y(t) + 1)^3, \qquad y(0) = 0, \qquad t \in T.$$

$$(VII.3-15)$$

Let the space Y be the Banach space of continuously differentiable functions on T satisfying $y(0) = 0$ and with norm (chosen for convenience)

$$\|y\| = \max_{t \in T} \max \{|e^t y(t)|, \tfrac{1}{2}|e^t \dot{y}(t)|\}. \qquad (VII.3-16)$$

A bound on $|y(t)|$ can be obtained from the norm of (VII.3-16). First note that $|y(t)| \le e^{-t}\|y\|$. Also,

$$|y(t)| \le \int_0^t |\dot{y}(s)|\, ds \le \int_0^t 2\|y\|e^{-s}\, ds = 2(1 - e^{-t})\|y\|. \qquad (VII.3-17)$$

Hence,

$$\max_{t \in T} |y(t)| \le \max_{t \in T} \min\{e^{-t}, 2(1 - e^{-t})\}\|y\| = \tfrac{2}{3}\|y\|. \qquad (VII.3-18)$$

Let Z be the Banach space of continuous functions with norm

$$\|z\| = \max_{t \in T} |e^t z(t)|. \qquad (VII.3-19)$$

X is the real line with absolute value for a norm. From (VII.3-5) the operator $z = P'_y(x_0, y_0)y$ is defined by

$$z(t) = \dot{y}(t) + y(t), \qquad t \in T. \qquad (VII.3-20)$$

The inverse operator $y = [P'_y(x_0, y_0)]^{-1} z$ is defined by [see VII.3-6)]

$$y(t) = \int_0^t e^{-(t-s)} z(s)\, ds, \qquad t \in T. \qquad (VII.3-21)$$

To bound $\|[P'_y(x_0, y_0)]^{-1}\|$ first note that from (VII.3-21), we have

$$|e^t y(t)| \leq \int_0^t |e^s z(s)| \, ds \leq \|z\|, \qquad t \in T. \qquad \text{(VII.3-22)}$$

From (VII.3-20) and (VII.3-22),

$$|e^t \dot{y}(t)| \leq |e^t z(t)| + |e^t y(t)| \leq 2\|z\|, \qquad t \in T. \qquad \text{(VII.3-23)}$$

Thus, from (VII.3-16) and the definition of a norm of a linear operator [see (I.4-5)],

$$\|[P_y'(x_0, y_0)]^{-1}\| \leq 1, \qquad \text{(VII.3-24)}$$

where the last norm is the one for $L(Z, Y)$.

Now the two nondecreasing functions for Theorem VII.2.1 must be found. With $\bar{y} \in Y$ and $t \in T$

$$e^t([P_y'(x, y) - P_y'(x_0, y_0)]\bar{y})(t) = e^t 3x[(y(t) + 1)]^2 \bar{y}(t)$$
$$= 3x[e^t(y(t) + 1)]^2 e^{-t} \bar{y}(t).$$
$$\text{(VII.3-25)}$$

Hence,

$$\|P_y'(x, y) - P_y'(x_0, y_0)\| \leq 3|x| \max_{t \in T} \left[e^t(y_0(t) + 1) + e^t(y(t) - y_0(t)) \right]^2 \tfrac{4}{9}$$
$$\leq \tfrac{4}{3} |x - x_0|(1 + \|y - y_0\|)^2 \qquad (x_0 = 0)$$
$$\equiv g_1(|x - x_0|, \|y - y_0\|). \qquad \text{(VII.3-26)}^\dagger$$

Also,

$$P(x, y_0)(t) = x(y_0(t) + 1)^3 = xe^{-3t}, \qquad t \in T, \qquad \text{(VII.3-27)}$$

so that

$$\|P(x, y_0)\| = \max_{t \in T} |xe^{-2t}| \leq |x - x_0| \equiv g_2(x - x_0). \qquad \text{(VII.3-28)}$$

Condition (7) of Theorem VII.2.1 becomes

$$\left\{ \begin{matrix} (\tfrac{4}{3})\delta(1 + \epsilon)^2 \leq \gamma < 1 \\ \delta \leq \epsilon(1 - \gamma) \end{matrix} \right\}. \qquad \text{(VII.3-29)}$$

For $\delta = 0.01$, it is found that (VII.3-29) is satisfied with $\gamma = 0.0136$, $\epsilon = 0.0102$. That is, if $|x| \leq 0.01$, the theorem states that $\|y - y_0\| \leq 0.0102$.

† $\max_{t \in T} e^{-t}|\bar{y}(t)| \leq \max_{t \in T} \min\{e^{-2t}, 2e^{-t}(1 - e^{-t})\} \|\bar{y}\| = \tfrac{4}{9} \|\bar{y}\|$

From the actual solution,† one finds that $\|y - y_0\| \leq 0.00432$. Note that the exact information is typically not available.

To use (VII.2-3), we have

$$\|P'_x(x, y) - P'_x(x_0, y_0)\|$$
$$\leq \max_{t \in T} e^{-t}|y(t) - y_0(t)| \cdot e^{2t} \cdot \left[(y(t) + 1)^2 + |y(t) + 1| \cdot |y_0(t) + 1|\right.$$
$$\left. + (y_0(t) + 1)^2\right]$$
$$\leq \tfrac{4}{9} \|y - y_0\|[(1 + \|y - y_0\|)^2 + (1 + \|y - y_0\|) + 1]. \qquad \text{(VII.3-30)}\ddagger$$

Relation (VII.2-3) becomes [using (VII.3-30) and (VII.3-26)]

$$\|y - y_0 - F'(x_0)(x - x_0)\|$$
$$\leq \sup_{\substack{|x-x_0|<\delta_1 \\ \|y-y_0\|\leq\epsilon_1}} \tfrac{4}{9} \|y - y_0\|[(1 + \|y - y_0\|)^2 + (1 + \|y - y_0\|) + 1]|x - x_0|$$
$$+ \sup_{\substack{|x-x_0|<\delta_2 \\ \|y-y_0\|\leq\epsilon_2}} \tfrac{4}{3} |x - x_0|(1 + \|y - y_0\|)^2\|y - y_0\|$$
$$= 1.35(10)^{-4}. \qquad \text{(VII.3-31)}$$

The actual bound on $\|y - y_0 - F'(x_0)(x - x_0)\|$ [using the norm of (VII.3-16)] is $0.281(10)^{-4}$. We could improve (VII.3-31) if we used the actual $\epsilon(=0.00432)$, but this information is usually not available (which is, of course, the motivation of this analysis). With the actual bounds unavailable (as is usually the case) the analysis might be interpreted to show the linearization to be accurate in the sense that the error in linearization ($\|y - y_0 - F'(x_0)(x - x_0)\|$) is substantially smaller than the changes in the solution (measured by ϵ).

Exercise VII.3.1

Indicate how to do the above example with the integral formulation

$$P(x, y)(t) = y(t) - \int_0^t f(y(s), x)\, ds \qquad \text{(VII.3-32)}$$

and compare the approaches.

Exercise VII.3.2

The above example addresses itself to parameter changes. How are initial condition changes handled? In particular, fill in the details below:

† The solution to (VII.3-12) is
$$y(t) = e^{-t}\,[1 + x(1 - e^{-2t})]^{-1/2} - 1, \qquad t \in T.$$

‡ Using $a^3 - b^3 = (a - b)(a^2 + ab + b^2)$,
$$[y(t) + 1]^3 - [y_0(t) + 1]^3 = (y(t) - y_0(t))[(y(t) + 1)^2 + (y(t) + 1) \cdot (y_0(t) + 1)$$
$$+ (y_0(t) + 1)^2].$$

$$\dot{y}(t) = f(y(t), t), \qquad y(0) = x,$$

Z is a space of continuously differentiable functions

$$P(x, y)(t) = y(t) - x - \int_0^t f(y(s), s) \, ds$$

$$(P_y'(x, y)\bar{y})(t) = \bar{y}(t) - \int_0^t f_y(y(s), s)\bar{y}(s) \, ds = z(t)$$

$$([P_y'(x, y)]^{-1}z)(t) = \Phi(t, 0)z(0) + \int_0^t \Phi(t, s)\dot{z}(s) \, ds$$

$$[F'(x_0)x](t) = \Phi(t, 0)x$$

$$\|\Gamma\| \le \max_t \left[\|\Phi(t, 0)\| + \int_0^t \|\Phi(t, s)\| \, ds \right]$$

$$\|y - y_0 - F'(x_0)(x - x_0)\| \le \|\Gamma\|\epsilon \, g_1(\delta, \epsilon).$$

Indicate how to use a second derivative with the last bound.

VII.4 RECONSIDERATION OF FIXED POINTS

By letting $P(x, y) = y - x(y)$ with x an operator, it is seen that finding a fixed point of x is a special case of finding a zero of $P(x, y)$. Hence, finding a fixed point of the operator x near a fixed point of the mapping x_0 could be considered as a special case of an implicit function problem. It may be noted that an arbitrary operator may not obviously be a member of a Banach space, an assumption placed on x in Sec. VII.2, but there are more general versions of implicit function theorems which do not make this assumption (see Ehrmann). Furthermore, we have been largely concerned with the following type of fixed point problem: given a $y_0 \in Y$, a Banach space, satisfying $y_0 = L_0 N(y_0)$, is there a $y \in Y$ satisfying $y = LN(y)$, where N is a fixed nonlinear operator mapping Y into itself and the bounded linear operators L and L_0 play the role of x and x_0? In this case, x and $x_0 \in L(Y, Y)$, the Banach space of bounded linear operators mapping Y into itself.

To discuss the implicit function approach to the fixed point problem just stated, assume N has a derivative at all $y \in Y$ and let us try to apply Theorem VII.2.1. We are assuming $x \in L(Y, Y)$ and

$$P(x, y) = y - xN(y), \tag{VII.4-1}$$

$$P(x_0, y_0) = 0. \tag{VII.4-2}$$

First,

$$P'_y(x_0, y_0) = I - x_0 N'(y_0) \tag{VII.4-3}$$

and it is required that this partial derivative have a bounded linear inverse. We know from Theorem I.4.3 that a sufficient condition for the existence of this inverse is that

$$\|x_0 N'(y_0)\| < 1. \tag{VII.4-4}$$

It should be recognized that (VII.4-4) is a condition that must be satisfied if the contraction mapping analyses of previous chapters are to be successful. Note, however, that (VII.4-4) is not a necessary condition for existence of the inverse of $P'_y(x_0, y_0)$ although it may be difficult in some cases to verify such existence without (VII.4-4). But the discussion in connection with (III.4-5) and (III.4-20) provides an example of the existence of an inverse without satisfying the conditions of Theorem I.4.3. Hence, it may be possible to satisfy the conditions of the implicit function theorem without having $xN(y)$ be a contraction near (x_0, y_0).

The above discussion is not completely satisfactory because Chapter II encouraged modifying mappings whose fixed points are sought so $\|x_0 N'(y_0)\| \geq 1$ does not necessarily put the CMT out of business. In fact, if we look back at the proof of Theorem VII.2.1 specialized to the case of $P(x, y) = y - xN(y)$, we see that we are essentially applying the CMT to a modification of the mapping xN. In particular, it is shown that $Q^{(x)}(y)$ is a contraction with

$$\begin{aligned} Q^{(x)}(y) &= y - \left[P'_y(x_0, y_0) \right]^{-1} P(x, y) \\ &= y - \left[I - x_0 N'(y_0) \right]^{-1} \left[y - xN(y) \right]. \end{aligned} \tag{VII.4-5}$$

If y is a fixed point of $Q^{(x)}(y)$, it is also a fixed point of $xN(y)$. This modification of $xN(y)$ is particularly attractive because

$$Q^{(x_0)'}(y_0) = 0. \tag{VII.4-6}$$

There is a slight difference between the implicit function and fixed point viewpoints. Implicit function theorems are usually concerned with all x such that $\|x - x_0\| < \delta$, while a fixed point theorem may be concerned only with a specific operator x. If interest is in a specific x, Newton's method is applicable which we now briefly discuss.

Having the derivative of a mapping be zero at an initial point is also a motivation for the *modified Newton's method* of finding a zero of a differentiable operator P defined on some Banach space, i.e., a y such that $P(y) = 0$. If Γ is an invertible linear operator then a fixed point of $y - \Gamma P(y)$ is also a zero of $P(y)$. What should Γ be? If the CMT is to be used to prove convergence, we should like to choose it so that $\|I - \Gamma P'(y)\|$ is as small as possible near y_0. A simple choice is $\Gamma = \left[P'(y_0) \right]^{-1}$ which at least makes $\|I - \Gamma P'(y_0)\| = 0$.

The modified Newton's method actually uses the following iteration to find a zero of $P(y)$:

$$y_{i+1} = y_i - [P'(y_0)]^{-1}P(y_i). \qquad \text{(VII.4-7)}$$

Conditions for convergence can be found using the CMT and by other methods. *Newton's method* itself uses

$$y_{i+1} = y_i - [P'(y_i)]^{-1}P(y_i) \qquad \text{(VII.4-8)}$$

which has better convergence properties than (VII.4-7) but at the expense of evaluating an inverse at each step. The literature on Newton's method is vast; see, e.g., Kantorovich and Akilov, Saaty, or Collatz.

The modified Newton's method may be used as a method of reducing Lipschitz constants and thus perhaps to extend the applicability of a contraction mapping analysis. Unfortunately, there is difficulty in the use of Newton's method in the case of nontrivial autonomous oscillations where we have already had a failure with a contraction mapping analysis (see Sec. V.8). Letting L and N be the operators of Sec. V.8, and letting

$$P(y) = y - F(y) \qquad (F(y) = LN(y)), \qquad \text{(VII.4-9)}$$

we are led to investigating the invertibility of $I - F'(y)$, where I is the identity operator. Consider

$$z = P'(y^*)y = (I - F'(y^*))y. \qquad \text{(VII.4-10)}$$

with y^* the fixed point of F (i.e., y^* is a nonconstant periodic solution of an autonomous differential equation). If $y = 0$, then $z = 0$. But because $F'(y^*)$ is associated with the equation of first variation, there is also a nonzero $y, \dot{y}^*$, which results in $z = 0$. Hence $P'(y^*)$ is not one-one and thus not invertible.

VII.5 NOTES

The discussion of partial derivatives follows Kantorovich and Akilov, Chapter XVII, Sec. 4. Implicit function theorems in Banach spaces are discussed in Kantorovich and Akilov and in Dieudonné. Theorem VII.2.1 is from Holtzman [5]. There has been much other work on implicit function theorems, some more general (see, e.g., Ehrmann which also has further references; a short discussion of Ehrmann's paper is given in Saaty) but the approach given here is appealing from the point of view of simplicity of application. The analysis of Sec. VII.3 is based on Holtzman [6]. Appendix VII.B follows Kantorovich and Akilov, pp. 689 and 690.

APPENDIX VII.A

Derivation of (VII.2-3)

We first derive a property involving partial derivatives (Kantorovich and Akilov, p. 686). Let P'_x and P'_y exist in a convex neighborhood of (x_0, y_0). Then, using the triangle inequality,

$$\|P(x_0 + \Delta x, y_0 + \Delta y) - P(x_0, y_0) - P'_x(x_0, y_0)\Delta x - P'_y(x_0, y_0)\Delta y\|$$

$$\leq \|P(x_0 + \Delta x, y_0 + \Delta y) - P(x_0, y_0 + \Delta y) - P'_x(x_0, y_0)\Delta x\|$$

$$+ \|P(x_0, y_0 + \Delta y) - P(x_0, y_0) - P'_y(x_0, y_0)\Delta y\|$$

$$\leq \sup_{0<\theta<1} \|P'_x(x_0 + \theta\Delta x, y_0 + \Delta y) - P'_x(x_0, y_0)\| \cdot \|\Delta x\|$$

$$+ \sup_{0<\theta_1<1} \|P'_y(x_0, y_0 + \theta_1\Delta y) - P'_y(x_0, y_0)\| \cdot \|\Delta y\|. \qquad \text{(VII.A-1)}$$

Hence,

$$\|P(x_0 + \Delta x, y_0 + \Delta y) - P(x_0, y_0) - P'_x(x_0, y_0)\Delta x - P'_y(x_0, y_0)\Delta y\|$$

$$\leq \min\{a, b\} \qquad \text{(VII.A-2)}$$

where a is the last quantity in (VII.A-1) and

$$b = \sup_{0<\theta<1} \|P'_x(x_0 + \theta\Delta x, y_0) - P'_x(x_0, y_0)\| \cdot \|\Delta x\|$$
$$+ \sup_{0<\theta_1<1} \|P'_y(x_0 + \Delta x, y_0 + \theta_1\Delta y) - P'_y(x_0, y_0)\| \cdot \|\Delta y\|.$$

$$(\text{VII.A-3})$$

Going from (VII.A-1) to (VII.A-2) merely involves adding and subtracting $P(x_0 + \Delta x, y_0)$ instead of $P(x_0, y_0 + \Delta y)$ in the manipulations of (VII.A-1). Going from the first to second inequality of (VII.A-1) uses the mean value formula (II.2-3) for the operator $G(x) = P(x, y_0 + \Delta y) - P'_x(x_0, y_0)x$, i.e., with $x = x_0 + \Delta x$,

$$\|G(x) - G(x_0)\| = \|P(x_0 + \Delta x, y_0 + \Delta y) - P(x_0, y_0 + \Delta y)$$
$$- P'_x(x_0, y_0)\Delta x\|$$
$$\le \sup_{\theta \in (0,1)} \|G'(x_0 + \theta\Delta x)\| \cdot \|\Delta x\|$$
$$= \sup_{\theta \in (0,1)} \|P'_x(x_0 + \theta\Delta x, y_0 + \Delta y) - P'_x(x_0, y_0)\| \cdot \|\Delta x\|.$$

$$(\text{VII.A-4})$$

Now, with $\Gamma = [P'_y(x_0, y_0)]^{-1}$, and recognizing that $P(x, y) = P(x_0, y_0) = 0$ (for δ sufficiently small),

$$\|y - y_0 - F'(x_0)(x - x_0)\|$$
$$= \|y - y_0 + \Gamma P'_x(x_0, y_0)(x - x_0)\|$$
$$= \|\Gamma[P'_x(x_0, y_0)(x - x_0) + P'_y(x_0, y_0)(y - y_0)]\|$$
$$\le \|\Gamma\| \cdot \|P(x, y) - P(x_0, y_0) - P'_x(x_0, y_0)(x - x_0)$$
$$- P'_y(x_0, y_0)(y - y_0)\|$$
$$\le \|\Gamma\| \min\{a, b\}. \qquad (\text{VII.A-5})$$

Derivation of $F'(x_0) = -\Gamma P'_x(x_0, y_0)$

To derive (VII.2-1) we must show that for any $\epsilon > 0$ there is a $\delta > 0$ such that $x - x_0 \in X$ and $||x - x_0|| < \delta$ implies

$$||F(x) - F(x_0) - F'(x_0)(x - x_0)|| \leq \epsilon||x - x_0|| \qquad \text{(VII.B-1)}$$

with $F'(x_0) = - \Gamma P'_x(x_0, y_0)$. Now,

$$
\begin{aligned}
F(x) &- F(x_0) - F'(x_0)(x - x_0) \\
&= y - y_0 + \Gamma P'_x(x_0, y_0)(x - x_0) \\
&= \Gamma[P'_x(x_0, y_0)(x - x_0) + P'_y(x_0, y_0)(y - y_0)]. \qquad \text{(VII.B-2)}
\end{aligned}
$$

Using $P(x, y) = P(x_0, y_0) = 0$, (VII.A-1) and with $(x - x_0) = \Delta x$, $(y - y_0) = \Delta y$, we have, for δ sufficiently small,

$$\|F(x) - F(x_0) - F'(x_0)\Delta x\| \leq \|\Gamma\| \cdot \| P(x, y) - P(x_0, y_0) - P'_x(x_0, y_0)\Delta x$$
$$- P'_y(x_0, y_0)\Delta y\|$$
$$\leq \|\Gamma\| \cdot [\sup_{0<\theta<1} \|P'_x(x_0 + \theta\Delta x, y_0 + \Delta y)$$
$$- P'_x(x_0, y_0)\| \cdot \|\Delta x\|$$
$$+ \sup_{0<\theta_1<1} \|P'_y(x_0, y_0 + \theta_1\Delta y)$$
$$- P'_y(x_0, y_0)\| \cdot \|\Delta y\|]$$
$$\leq \eta[\|\Delta x\| + \|\Delta y\|] \qquad \text{(VII.B-3)}$$

where η can be made small as desired by choosing δ sufficiently small since P'_x and P'_y are both assumed continuous and $\|\Delta y\| \to 0$. Now,

$$\|F(x) - F(x_0) - F'(x_0)\Delta x\| \leq \eta[\|x - x_0\| + \|F(x) - F(x_0)\|]$$
$$\leq \eta[\|x - x_0\| + \|F'(x_0)\Delta x\|$$
$$+ \|F(x) - F(x_0) - F'(x_0)\Delta x\|].$$
$$\text{(VII.B-4)}$$

If $\eta < 1$,

$$\|F(x) - F(x_0) - F'(x_0)\Delta x\| \leq \frac{\eta[\|x - x_0\| + \|F'(x_0)\Delta x\|]}{1 - \eta}$$
$$\leq \frac{\eta[1 + \|F'(x_0)\|]\|\Delta x\|}{1 - \eta}. \qquad \text{(VII.B-5)}$$

By choosing δ so small that

$$\frac{\eta[1 + \|F'(x_0)\|]}{1 - \eta} \leq \epsilon, \qquad \text{(VII.B-6)}$$

the derivation is completed.

VIII

Stability of Nonlinear Feedback Systems

The results of Chapters III and IV focused attention on responses of nonlinear feedback loops on $(-\infty, \infty)$. In Sec. III. 4, the input and output are in $L_2(-\infty, \infty)$; in Chapter IV, input and output are periodic functions which are defined on $(-\infty, \infty)$. In Secs. III. 5 and IV. 5, it was specifically pointed out that initial conditions and transient effects were not being explicitly considered, i. e., steady-state response was being emphasized. In this chapter we return to the study of nonlinear feedback loops with an interest in initial condition and transient effects. For analysis of these properties, we study the feedback loop's response on $[0, \infty)$ with the system having some initial condition at $t = 0$. Our objective is to obtain frequency domain conditions which imply certain properties of the response. We shall reconsider Aizerman's conjecture (already discussed in Sec. VI. 4) in the light of the stability conditions derived.

The analysis of Chapter III will be seen to be also relevant in this chapter. However, in view of the fact that the conditions of Chapter III are quite stringent, we shall see how to relax the requirement (not require a contraction) and get less restrictive conditions.

VIII.1 GENERAL COMMENTS

There are many definitions of stability. We first give here the familiar definitions of Lyapunov stability for differential equations. The solution x of the vector differential equation

$$\frac{dx(t)}{dt} = f(x(t), t), \qquad t \geq 0, \qquad \text{(VIII. 1-1)}$$

is *Lyapunov stable* (or stable) if for all $t_0 \geq 0$ and all $\epsilon > 0$, there is a $\delta(\epsilon, t_0)$ such that

$$|x(t_0) - \tilde{x}(t_0)| < \delta \Rightarrow \tilde{x}(t) \text{ exists on } [t_0, \infty)$$

and

$$|x(t) - \tilde{x}(t)| < \epsilon, \qquad t \geq t_0, \qquad \text{(VIII. 1-2)}$$

with the vertical bars denoting magnitude of the vector and with $\tilde{x}(t)$ a solution of (VIII. 1-1) with initial condition $\tilde{x}(t_0)$ at $t = t_0$. The solution x of (VIII. 1-1) is *asymptotically stable* if it is stable and if there is a $\delta(t_0) > 0$ such that

$$|x(t_0) - \tilde{x}(t_0)| < \delta \Rightarrow \lim_{t \to \infty} |x(t) - \tilde{x}(t)| = 0. \qquad \text{(VIII. 1-3)}$$

If in (VIII. 1-3) δ may be any positive number we say that the solution is *globally asymptotically stable*.

In (VIII. 1-2) and (VIII. 1-3), x is always compared with $\tilde{x}$ at the same time. Sometimes a system may have two solutions which actually resemble each other but they may be out of synchronism in time. This is particularly true of periodic solutions of autonomous systems. Hence, we make the following definition : the nonconstant periodic solution x of (VIII. 1-1) is *orbitally stable* if for all $\epsilon > 0$ there is a $\delta(\epsilon, t_0) >$ such that

$$d(\tilde{x}(t_0), C) < \delta \Rightarrow d(\tilde{x}(t), C) < \epsilon, \qquad t \geq t_0, \qquad \text{(VIII. 1-4)}$$

where C represents the closed orbit of x, i.e.,

$$d(\tilde{x}(t), C) = \inf\{|\tilde{x}(t) - x(t')| : t' \geq 0\}. \qquad \text{(VIII. 1-5)}$$

[We are assuming existence of $\tilde{x}(t)$]. The periodic solution x is *orbitally asymptotically stable* if it is orbitally stable and, in addition, there is a $\delta(t_0) > 0$ such that

$$d(\tilde{x}(t_0), C) < \delta \Rightarrow \lim_{t \to \infty} d(\tilde{x}(t), C) = 0. \qquad \text{(VIII. 1-6)}$$

In Sec. VI. 3, we discussed the periodic solutions of the autonomous Duffing's and Van der Pol's differential equations. The reader should relate the orbital stability definitions to the phase-plane plots of Figure VI.1.

We shall not actually use the preceding widely used stability definitions. Rather we shall use some definitions better suited to input-output analysis of feedback systems. The preceding definitions, however, show that there is a connotation of continuity. The orbital stability definition shows this most clearly with the use of metrics in (VIII. 1-4). With (VIII. 1-2), we might consider the mapping from initial condition $x(0)$ to x, a solution of the differential equation,† in a space of continuous functions with $\|x\| = \sup\limits_{t \geq 0} |x(t)|$. Then (VIII. 1-2) implies a continuity property. Stability requires that solutions can be made to be as close as desired by making the initial conditions close enough.

The same type of stability definitions may be made for nonlinear feedback systems except that we have as much interest in inputs as in initial conditions. In fact, we find it convenient to lump the initial conditions effects in with the input and then (loosely speaking) the feedback system is called continuous if changes in the output can be made small by making changes in the input small.

The feedback system to be considered is shown in Figure VIII.1. This is the same feedback system as considered in Chapter III and IV except that in this

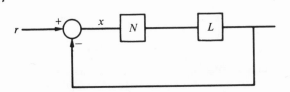

Figure VIII.1 Nonlinear feedback loop

chapter, we are specifically interested in initial condition and transient effects while the previous chapters emphasized steady-state behavior. An important special case of the system to be studied occurs when the box L in Figure VIII.1 can be represented by the following:‡

$$z(t) = z_0(t) + \int_0^t h(t - u)y(u)\,du, \qquad \text{(VIII. 1-7)}$$

where $z_0(t)$ is a function taking initial conditions into account (see, e.g., Zadeh and Desoer, p. 233) and $y = Nx$ is given by

† According to our definition of mapping in Sec. I. 4, we require a mapping to be single-valued. Hence, in this informal discussion we are assuming existence of *unique* solutions, an assumption not always made in stability analyses.

‡ At this point, then, the box labelled L represents a linear operator and an additive term.

$$y(t) = n(x(t), t).$$ (VIII. 1-8)

The equation describing the complete feedback loop in Figure VIII.1 is

$$x(t) = r(t) - z_0(t) - \int_0^t h(t - u)n(x(u), u)\, du$$ (VIII. 1-9)

which shows that the initial condition function just adds to the input. From now on, it is assumed that r includes the initial condition function. Assumptions made on r will be passed along to any initial condition functions implicitly considered.

Sometimes (as in Sec. VI. 4), the order of L and N is interchanged from that shown in Figure VIII.1. If the feedback system is as in Figure VIII. 2, the equation representing the system is

$$y = - N(Ly + z_0) + q.$$ (VIII. 1-10)

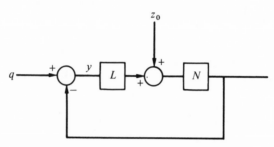

Figure VIII.2 Feedback loop with L and N reversed

This may be transformed to $x = - LNx + r$ by making the following identifications :

$$x = Ly + z_0, \qquad r = Lq + z_0.$$ (VIII. 1-11)

The general viewpoint taken here towards stability concerns itself with properties of the system's responses to given inputs. To informally discuss this, let x be defined on $[0, \infty)$ and define a truncation operator as follows :

$$(P_T x)(t) = x(t) \qquad t \in [0, T]$$
$$= 0 \qquad t > T.$$ (VIII. 1-12)

Define

$$L_{2e}(0, \infty) = \{x : P_T x \in L_2(0, \infty) \qquad \text{for all } T \in (0, \infty)\}.$$ (VIII. 1-13)

If $x \in L_{2e}(0, \infty)$, then all of its truncations as given in (VIII. 1-12) are in $L_2(0, \infty)$. This does not, of course, imply that $x \in L_2(0, \infty)$; e.g., if $x(t) = e^t$ then

$x \in L_{2e}(0, \infty)$ but $x \notin L_2(0, \infty)$. With the system described by $x = -LNx + r$ it is assumed that $x \in L_{2e}(0, \infty)$. We thereby divorce the question of existence of solutions from the problem of stability (defined in terms of properties of solutions). They can often be answered by separate arguments. The existence question might be resolved by a result such as Example II. 1.2. Also, there are standard existence theorems for integral equations in, e.g., Tricomi. At any rate, we use $x \in L_{2e}(0, \infty)$ as a starting point. We emphasize $x \in L_{2e}(0, \infty)$ is a weak condition but to have $x \in L_2(0, \infty)$ is a relatively strong requirement.

If $x = -LNx + r$, $x \in L_{2e}(0, \infty)$ and for all $r \in L_2(0, \infty)$ we have $x \in L_2(0, \infty)$ and

$$\|x\| \leq k_1 \|r\|, \qquad k_1 < \infty, \qquad \text{(VIII. 1-14)}$$

then we say that the responses are (or the system is) *bounded*. Note that we are not just proving the existence of an $x \in L_2(0, \infty)$ satisfying $x = -LNx + r$. This would still allow the existence of another $x \in L_{2e}(0, \infty)$, but which is not in $L_2(0, \infty)$, satisfying the equation. We want to show that *all* $x \in L_{2e}(0, \infty)$ satisfying the equation must be in $L_2(0, \infty)$. This stability definition does not have the continuity connotation of Liapunov stability.

If $x_1, x_2 \in L_{2e}(0, \infty)$, the system is called *continuous* if for any $r_1 - r_2 \in L_2(0, \infty)$, the respective responses satisfy $x_1 - x_2 \in L_2(0, \infty)$ and

$$\|x_1 - x_2\| \leq k_2 \|r_1 - r_2\|, \qquad k_2 < \infty. \qquad \text{(VIII. 1-15)}$$

This definition is actually stronger than the straightforward adaptation of the definition of continuity of a function which would require that for any $\epsilon > 0$ there is a $\delta(\epsilon) > 0$ such that

$$\|r_1 - r_2\| < \delta \Rightarrow \|x_1 - x_2\| < \epsilon. \qquad \text{(VIII. 1-16)}$$

We use the stronger definition because the methods used lead to that type of result. The method of deriving conditions for continuity will be reminiscent of a contraction mapping analysis although our motivation is slightly different [we are assuming existence in $L_{2e}(0, \infty)$, not proving existence as we do in the usual contraction mapping application]. However, under slight additional conditions, existence in $L_{2e}(0, \infty)$ will also be implied. On the other hand, the conditions for boundedness will not be shown here to imply existence in $L_{2e}(0, \infty)$.

Though our emphasis is on the case of $L_2(0, \infty)$ the basic propositions are derived for more general normed linear spaces and inner product spaces. This is done not merely for the sake of generality but the results actually become quite transparent in their generality (they may actually be even further generalized but the form to be given suffices for our purposes). We now proceed to two abstract theorems on boundedness and continuity.

VIII.2 TWO ABSTRACT THEOREMS ON
BOUNDEDNESS AND CONTINUITY

Let Ω be a fixed nonempty set of real numbers. If $T \in \Omega$, denote $[T] = (- \infty, T) \cap \Omega$. Let $\mathscr{L}$ be be a linear space and $\tilde{X}$ be the linear space of all mappings from Ω into $\mathscr{L}$ (with customary addition and scalar multiplication). Let X_e and X be linear subspaces of $\tilde{X}$ such that $X \subset X_e \subset \tilde{X}$. Also, let X be a normed linear space. X_e is called the extension of X, the nomenclature following from the following relationship between X and X_e [also look back at (VIII. 1-13)].

If $x \in \tilde{X}$, define the linear truncation operator P_T as follows :

$$(P_T x)(t) = x(t), \qquad t \in [T]$$
$$= 0, \qquad t \in \Omega, \; t > T. \qquad \text{(VIII. 2-1)}$$

For brevity, let x_T denote $P_T x$. It is assumed that if, for $x \in \tilde{X}$, we have $x_T \in X$ for all $T \in \Omega$, then $x \in X_e$. Furthermore, it is assumed that $x \in X$ with $\|x\| \leq M < \infty$ if $x \in X_e$ and $\|x_T\| \leq M$ for all $T \in \Omega$.

The following properties of the operator P_T will be used. The first is an assumption and the other two are clear.

(a)† If $x \in X$ and $T \in \Omega$, then $\|x_T\| \leq \|x\|$.   (VIII. 2-2)

(b)   If $x, y \in X_e$ and $T \in \Omega$, then $x(t) = y(t)$ on $[T]$

     if and only if $x_T = y_T$.   (VIII. 2-3)

(c) $P_{T_1} P_{T_2} = P_{T_1}$ if $T_1 \leq T_2$ and $T_1, T_2 \in \Omega$.   (VIII. 2-4)

The operators P_T play an important role in what is to follow. From (VIII. 2-4) it is seen that $P_T^2 = P_T$. Exercise IV. 1.2 showed that, if P is a projection, then $P^2 = P$. Conversely, it may be shown that if a linear operator P satisfies $P^2 = P$, then P is a projection. In fact, some authors define projections by means of the property $P^2 = P$ (see Porter, p. 109). Hence, P_T is a projection operator on X_e.

We say that a (not necessarily linear) operator $F \in \Theta$ if and only if F maps X_e into itself and $P_T F = P_T F P_T$ on X_e for all $T \in \Omega$. In other words, $(Fx)_T = (Fx_T)_T$ for all $x \in X_e$, all $T \in \Omega$. An operator in Θ is called *nonanticipative* since $(Fx)(t)$ does not depend on future values of $x(t)$.

Exercise VIII.2.1

Show that $F_1, F_2 \in \Theta \Rightarrow F_1 F_2 \in \Theta$ and $F_1 + F_2 \in \Theta$. Give an example of an $F \in \Theta$ for which $F^{-1} \notin \Theta$.

---

†   This is an assumption, not a consequence of the definition of P_T. A counterexample to (a) being generally true may be obtained by using a space of functions of bounded variation (Kantorovich and Akilov, p. 74) and letting x be a nonzero constant.

As examples of the above definitions, let $\Omega = [0, \infty)$ and $\mathscr{L} = (-\infty, \infty)$. Then $\tilde{X}$ is the space of all real-valued functions defined on $[0, \infty)$. Let

$$X_e = \left\{ x : x \in \tilde{X}, \quad x \text{ measurable}, \quad \int_0^T x^2(t)\, dt < \infty \quad \text{for all } T \in (0, \infty) \right\}$$

$$= L_{2e}(0, \infty), \tag{VIII. 2-5}$$

$$X = \left\{ x : x \in \tilde{X}, \quad x \text{ measurable}, \quad \int_0^\infty x^2(t)\, dt < \infty \right\} = L_2(0, \infty).$$

$$\tag{VIII. 2-6}$$

The following operator F is in Θ,

$$(Fx)(t) = \int_0^\infty h(t - u)x(u)\, du, \qquad x \in X_e, \tag{VIII. 2-7}$$

if $h \in L_1(0, \infty)$ and $h(t) = 0$ for $t < 0$, i.e.,

$$(Fx)(t) = \int_0^t h(t - u)x(u)\, du. \tag{VIII. 2-8}$$

The reader should verify that the above examples are indeed consistent with all of the preceding definitions.

It should be recognized that the results to be given are much more generally applicable than just to $L_{2e}(0, \infty)$. For example, different cases of Ω give rise to sampled-data systems, continuous-time systems, systems operating over finite or infinite intervals, etc.

Let F map X_e into itself and $F(0) = 0$. The *gain* of F, $g(F)$, is defined as

$$g(F) = \sup \left\{ \frac{\|(Fx)_T\|}{\|x_T\|} : x \in X_e, \quad T \in \Omega, \quad x_T \neq 0 \right\}. \tag{VIII. 2-9}$$

The formula for gain resembles that for the norm of a linear operator, but F is not assumed to be linear. If F maps X_e into itself, the *incremental gain* of F, $g_i(F)$, is

$$g_i(F) = \sup \left\{ \frac{\|(Fx_1 - Fx_2)_T\|}{\|(x_1 - x_2)_T\|} : x_1, x_2 \in X_e, \quad T \in \Omega, \quad (x_1 - x_2)_T \neq 0 \right\}. \tag{VIII. 2-10}$$

The incremental gain is similar to a smallest Lipschitz constant.

Exercise VIII.2.2

Suppose F maps X_e into itself and $g(F) < \infty$. Does this imply that $F \in \Theta$? Consider $(Fx)(t) = \inf \{ |x(\tau)| : \tau \in [0, t + t_1] \}, t_1 > 0$.

Now consider the system described by

$$x = -LNx + r \qquad \text{(VIII. 2-11)}$$

for which we have the following two theorems, the first leading to bounded-ness and the second leading to continuity. We generalize the definitions of boundedness and continuity given in connection with (VIII. 1-14) and (VIII. 1-15) by replacing $L_2(0, \infty)$ and $L_{2e}(0, \infty)$ with X and X_e, respectively.

Theorem VIII.2.1

Let $L \in \Theta$ and let L be a linear map of X_e into itself. Let N map X_e into itself and $N(0) = 0$. Let $x \in X_e$ and $x = -LNx + r$. Suppose there is a λ such that

(i) $(I + \lambda L)$ is invertible on X_e, $(I + \lambda L)^{-1} \in \Theta$, and $L(I + \lambda L)^{-1}$ is a bounded linear map of X into itself,

(ii) with $\eta_\lambda = \|L(I + \lambda L)^{-1}\|$ and $k_\lambda = g(N - \lambda I)$,

$$\eta_\lambda k_\lambda < 1. \qquad \text{(VIII. 2-12)}$$

Then

$$\|x_T\| \le (1 - \eta_\lambda k_\lambda)^{-1} (1 + |\lambda|\eta_\lambda)\|r_T\| \qquad \text{(VIII. 2-13)}$$

for all $T \in \Omega$.

Remark. The hypotheses imply that $r \in X_e$.

Corollary

If $r \in X$, then $x \in X$ and

$$\|x\| \le (1 - \eta_\lambda k_\lambda)^{-1}(1 + |\lambda|\eta_\lambda)\|r\|, \qquad \text{(VIII. 2-14)}$$

i. e., the system is bounded.

Proof. Let $T \in \Omega$. Then, with the following manipulations,

$$\lambda LNx = -\lambda x + \lambda r$$
$$(I + \lambda L)Nx = (N - \lambda I)x + \lambda r$$
$$Nx = (I + \lambda L)^{-1}[(N - \lambda I)x + \lambda r],$$

we have, using the nonanticipativeness of L and $(I + \lambda L)^{-1}$,

$$x_T = -P_T L(I + \lambda L)^{-1}[(N - \lambda I)x + \lambda r] + P_T r$$

$$= - P_T L(I + \lambda L)^{-1} P_T[(N - \lambda I)x + \lambda r] + r_T,$$

$$\|x_T\| \leq \eta_\lambda \|P_T[(N - \lambda I)x + \lambda r]\| + \|r_T\|$$

$$\leq \eta_\lambda \|P_T(N - \lambda I)x\| + \eta_\lambda |\lambda| \cdot \|r_T\| + \|r_T\|$$

$$\leq \eta_\lambda \, k_\lambda \|x_T\| + (1 + |\lambda| \eta_\lambda) \|r_T\|,$$

$$(1 - \eta_\lambda \, k_\lambda) \|x_T\| \leq (1 + |\lambda| \eta_\lambda) \|r_T\|.$$

Since $\eta_\lambda \, k_\lambda < 1$, we can divide through by $(1 - \eta_\lambda \, k_\lambda)$ to obtain the desired result.

Remarks. (i) The manipulation of the mappings in the proof is similar to, but not identical to, the manipulation in Sec. III. 2. The above manipulation does not require L to map X into itself. The significance of this will be discussed in Sec. VIII. 4.

(ii) In the above proof, we did not use all of the structure imposed on the mappings P_T. We used the following facts about P_T : (VIII. 2-2) and the definition of nonanticipativeness. The additional structure is often present in applications and is used in the discussion of existence in Appendix VIII. A. The corollary used the relationship between X and X_e.

Exercise VIII.2.3

Where are the following used in the proof of Theorem VIII. 2.1 : (VIII. 2-2) and $L(I + \lambda L)^{-1}$ maps X into itself ?

How can the theorem be modified to handle $N(0) \neq 0$?

Theorem VIII.2.2

Let $L \in \Theta$ and let L be a linear map of X_e into itself. Let N map X_e into itself. Let $x_1, x_2 \in X_e$ and

$$x_1 = - LN x_1 + r_1,$$
$$x_2 = - LN x_2 + r_2. \qquad \text{(VIII. 2-15)}$$

Suppose there is a λ such that

(i) $(I + \lambda L)$ is invertible on X_e, $(I + \lambda L)^{-1} \in \Theta$, and $L(I + \lambda L)^{-1}$ is a bounded linear map of X into itself.

(ii) with $\eta_\lambda = \|L(I + \lambda L)^{-1}\|$ and $k_\lambda = g_i(N - \lambda I)$,

$$\eta_\lambda \, k_\lambda < 1. \qquad \text{(VIII. 2-16)}$$

Then

$$\|P_T(x_1 - x_2)\| \le (1 - \eta_\lambda k_\lambda)^{-1}(1 + |\lambda|\eta_\lambda)\|P_T(r_1 - r_2)\| \qquad \text{(VIII. 2-17)}$$

for all $T \in \Omega$.

Corollary

If $r_1 - r_2 \in X$, then $x_1 - x_2 \in X$ and

$$\|x_1 - x_2\| \le (1 - \eta_\lambda k_\lambda)^{-1}(1 + |\lambda|\eta_\lambda)\|r_1 - r_2\|, \qquad \text{(VIII. 2-18)}$$

i. e., the system is continuous.

Proof. As in the proof of Theorem VIII. 2.1,

$$Nx_1 = (I + \lambda L)^{-1}[(N - \lambda I)x_1 + \lambda r_1]$$
$$Nx_2 = (I + \lambda L)^{-1}[(N - \lambda I)x_2 + \lambda r_2]$$

so that

$$
\begin{aligned}
P_T x_1 - P_T x_2 = {}& -P_T L(I + \lambda L)^{-1}[(N - \lambda I)x_1 + \lambda r_1 - (N - \lambda I)x_2 - \lambda r_2] \\
& + P_T r_1 - P_T r_2 \\
= {}& -P_T L(I + \lambda L)^{-1}P_T[Nx_1 - Nx_2 - \lambda(x_1 - x_2) + \lambda(r_1 - r_2)] \\
& + P_T(r_1 - r_2), \\
\|P_T(x_1 - x_2)\| \le {}& \eta_\lambda \|P_T[Nx_1 - Nx_2 - \lambda(x_1 - x_2)]\| + \eta_\lambda \|P_T \lambda(r_1 - r_2)\| \\
& + \|P_T(r_1 - r_2)\| \\
\le {}& \eta_\lambda k_\lambda \|P_T(x_1 - x_2)\| + (1 + |\lambda|\eta_\lambda)\|P_T(r_1 - r_2)\|.
\end{aligned}
$$

Remark. The above proof is reminiscent of the CMT. Though we have emphasized that the questions of existence and stability are being divorced, the hypotheses of the above theorem actually imply existence in X_e if some minor additional assumptions are made. That is, the existence assumption could be traded for other hypotheses. See Appendix VIII. A.

Our next task is to interpret the above conditions in a familiar space, namely, $L_2(0, \infty)$. We note at this point, however, the relationship of this section to Chapter III. In Theorems VIII. 2.1 and VIII. 2.2, it is required that

$$\eta_\lambda k_\lambda = \|L(I + \lambda L)^{-1}\|k_\lambda < 1. \qquad \text{(VIII. 2-19)}$$

It is recognized that this is the same type of requirement as given in Theorem III. 1.1 if k_λ is of the form of $\eta(c)$ given by (III. 1-3). An important application (given in the next section) of Theorems VIII. 2.1 and VIII.2.2 to the case of $X = L_2(0, \infty)$ leads to such a k_λ. Hence, according to Theorem III. 2.1, $\lambda = \frac{1}{2}(\alpha + \beta)$ minimizes $\eta_\lambda k_\lambda$. The reader can easily verify that the proof of Theorem III. 2.1 goes through with the order of L and $(I + cL)^{-1}$ interchanged.

VIII.3 INTERPRETATION IN $L_2(0, \infty)$

Our objective here is to apply the two abstract results of the last section in the setting of $L_2(0, \infty)$ with a view towards getting a circle condition interpretation related to that given in Sec. III. 4.

Let $X = L_2(0, \infty)$ and $X_e = L_{2e}(0, \infty)$ [see (VIII. 2-5) and (VIII. 2-6)]. The linear operator $y = Lx$ is defined by

$$y(t) = \int_0^t h(t - u)x(u)\,du \qquad \text{(VIII. 3-1)}$$

with $h \in L_1(0, \infty)$. We already know that the operator defined by (VIII. 3-1) maps $L_2(0, \infty)$ into itself. Theorems VIII. 2.1 and VIII. 2.2 require that L map X_e into itself [a fact we stated in connection with (VIII. 2-8)]. That $L_{2e}(0, \infty)$ is mapped into itself is shown easily as follows. If $x \in L_{2e}(0, \infty)$, then $x_T \in L_2(0, \infty)$ for all $T \epsilon (0, \infty)$. Since $(Lx)_T = (Lx_T)_T$ for all $T \in (0, \infty)$, L maps $L_{2e}(0, \infty)$ into itself and, furthermore, $L \in \Theta$.

We remark that we are making a stronger assumption here than is required for Theorems VIII. 2.1 and VIII. 2.2, namely that L maps $L_2(0, \infty)$ into itself rather than just mapping $L_{2e}(0, \infty)$ into itself. This is done at this point for simplicity; we shall further discuss this in Sec. VIII. 4.

The nonlinear operator $y = Nx$ is defined by

$$y(t) = n(x(t), t) \qquad \text{(VIII. 3-2)}$$

with $n(x(t), t)$ real-valued and measurable if $x(t)$ is, $n(0, t) = 0$, and one of the following two conditions are satisfied:

(A1) $\alpha \le \dfrac{n(u, t)}{u} \le \beta \; (\beta > 0)$ for all $t \in [0, \infty)$ and all real $u \ne 0$.

$$\text{(VIII. 3-3)}$$

(A2) $\alpha \le \dfrac{n(u_1, t) - n(u_2, t)}{u_1 - u_2} \le \beta \; (\beta > 0)$ for all $t \in [0, \infty)$ and all real u_1,

u_2 with $u_1 \ne u_2$.

$$\text{(VIII. 3-4)}$$

Conditions (A1) and (A2) will be used in applying Theorems VIII. 2.1 and VIII. 2.2, respectively. It should be clear that, under either condition, $N \in \Theta$. Furthermore, under condition (A1),

$$g(N - \lambda I) = \max\{|\beta - \lambda|, |\lambda - \alpha|\} \tag{VIII. 3-5}$$

and, under condition (A2),

$$g_i(N - \lambda I) = \max\{|\beta - \lambda|, |\lambda - \alpha|\}. \tag{VIII. 3-6}$$

Condition (A1) may be interpreted as follows. For any fixed $t \in [0, \infty)$, the graph of the function $n(\cdot, t)$ lies between two straight lines with slope α and β (see Figure. VIII. 3). Note that the nonlinearity need not be known exactly; we need only know that its graph lies in a given sector. This can be a strength or weakness of this type of analysis. From a practical point of view, we can never know the function exactly so we would like conditions not tied to too exact specifications. On the other hand, we might actually know more about the nonlinearity than it just lies in a sector (e. g., it is nondecreasing or odd or approximately odd in some sense). Without using the additional information available, the more general result may be unduly conservative. Indeed, much recent work is in exploiting special characteristics of the nonlinearity to sharpen stability conditions.

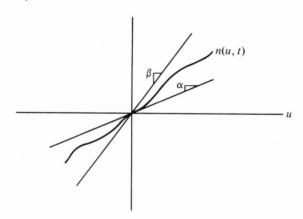

Figure VIII.3 Graph of $n(u, t)$ for fixed t

We will show that the conditions of Theorems VIII. 2.1 and VIII. 2.2 are satisfied if (A1) and (A2) are satisfied, respectively, and if the following are satisfied (for both theorems): with $s = \sigma + i\omega$ and

$$H(s) = \int_0^\infty h(t)e^{-st}\, dt \quad \text{for} \quad \sigma \geq 0 \tag{VIII. 3-7}$$

(i) $[1 + \tfrac{1}{2}(\alpha + \beta)H(s)] \neq 0 \quad \text{for} \quad \sigma \geq 0 \tag{VIII. 3-8}$

(ii) $\displaystyle\sup_{\omega \in (-\infty, \infty)} \left| \frac{H(i\omega)}{1 + \tfrac{1}{2}(\alpha + \beta)H(i\omega)} \right| \tfrac{1}{2}(\beta - \alpha) < 1. \tag{VIII. 3-9}$

$H(s)$ is, of course, the Laplace transform of h which may be defined as the Fourier transform of $h(t)e^{-\sigma t}$ with $h(t)e^{-\sigma t} \in L_1(0, \infty)$. However, we take advantage of the properties of $H(s)$ as a function of a complex variable. The above conditions should be compared with the conditions of Sec. III. 4; in particular, see (III.4-5) and (III.4-6). Condition (VIII.3-8) differs from (III.4-5) in that (VIII.3-8) must be satisfied for all s in the right half of the complex plane, not just on the imaginary axis. As mentioned in Sec. III. 5, this is a consequence of requiring $(I + \frac{1}{2}(\alpha + \beta)L)$ to be invertible in a different space, than that used in Sec. III. 4.

Just as in the proof of Nyquist's criterion for the rational case, (VIII.3-8) can be interpreted geometrically using the principle of the argument (an appropriate version is given in Evgrafov, pp. 97 and 98). This principle is concerned with a function $F(s)$ analytic in a domain D (an open connected† set in the complex plane) and continuous up to C, the boundary of D (analyticity is defined only at interior points), with the exception of a finite number of poles. The function $F(s)$ is assumed not to vanish on C nor become infinite there. The argument principle is that $N_0 - N_p = (2\pi)^{-1}\triangle \arg F(s)$, where N_0 and N_p are the number of zeros and poles, respectively, of $F(s)$ in D and $\triangle \arg F(s)$ is the change in the argument of $F(s)$ as s traverses D in the positive direction (such that D is on the left). That is, the number of times the locus of $F(s)$ winds around the origin equals $N_0 - N_p$.

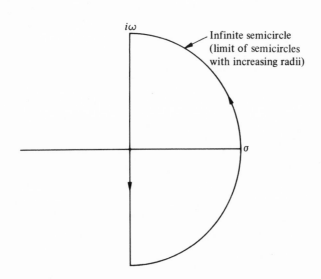

Figure VIII.4 Boundary of D

† An open set D in the complex plane is connected if any two of its points can be joined by a polygonal line all of whose points are in D (Evgrafov, p. 8).

In our case, let C be as shown in Figure VIII.4 and $F(s) = 1 + \frac{1}{2}(\alpha + \beta)H(s)$. It may be shown that $H(s)$ vanishes on C (for $s = \pm i\omega$ by the Riemann-Lebesgue lemma and for $s \to \infty$ with Re $s > 0$, see Doetsch, p. 190). Also, it may be shown that $H(s)$ is analytic in Re $s > 0$ and continuous up to the imaginary axis. Then (VIII. 3-8) becomes equivalent to the locus of $H(i\omega)$ for $\omega \in (-\infty, \infty)$ not encircling the point $(-2/(\alpha + \beta), 0)$. Hence (VIII. 3-8) and (VIII. 3-9) are interpreted together as follows:

(i) If $\alpha > 0$, the locus of $H(i\omega)$ for $\omega \in (-\infty, \infty)$ lies outside and does not encircle the circle C_1 shown in Figure VIII. 5(a).

(ii) If $\alpha = 0$, Re $[H(i\omega)] > -\beta^{-1}$ for $\omega \in (-\infty, \infty)$.

(iii) If $\alpha < 0$, the locus of $H(i\omega)$ for $\omega \in (-\infty, \infty)$ is contained within the circle C_2 shown in Figure VIII. 5(b).

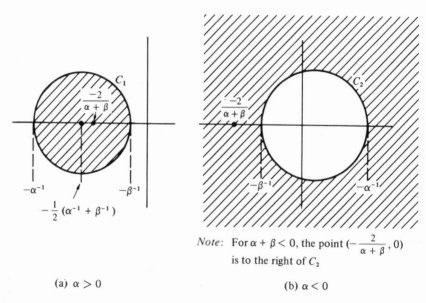

Note: For $\alpha + \beta < 0$, the point $(-\dfrac{2}{\alpha + \beta}, 0)$ is to the right of C_2

(a) $\alpha > 0$ (b) $\alpha < 0$

Figure VIII-5 Geometric interpretation of stability conditions

These are the same conditions as (i)–(iii) below (III. 4-6) except that in (i) there is an added restriction here about encirclement. For $\beta > 0$, observe that as $\beta - \alpha \to 0$, the circle collapses to a point and the circle condition becomes Nyquist's criterion (more precisely, the sufficient part of Nyquist's criterion).

We now show that the conditions of Theorems VIII. 2-1 and VIII. 2-2 are satisfied. For both theorems, it is required that $(I + \lambda L)$ be invertible on $X_e = L_{2e}(0, \infty)$ and $(I + \lambda L)^{-1} \in \Theta$. We show this via two lemmas. The first lemma makes use of the convention that functions that differ on zero-measure sets are considered to be the same (see the second footnote in Sec. III. 3).

Lemma 1

Let G be an invertible linear mapping of $L_2(0, \infty)$ onto itself such that for arbitrary $x \in L_2(0, \infty)$,

$$GS_\tau x = S_\tau Gx, \qquad \tau > 0, \tag{VIII. 3-10}$$

where S_τ is the mapping of $L_2(0, \infty)$ into itself defined by [for $x \in L_2(0, \infty)$]

$$\begin{aligned} S_\tau x(t) &= 0 \qquad t \in [0, \tau) \\ &= x(t - \tau) \qquad t \in [\tau, \infty). \end{aligned} \tag{VIII. 3-11}$$

Then, for all $x \in L_2(0, \infty)$,

$$P_T G^{-1} x = P_T G^{-1} P_T x \tag{VIII. 3-12}$$

Proof. Suppose, on the contrary, that G^{-1} does not satisfy (VIII. 3-12). Then there is a $T \in (0, \infty)$ and an $x_1 \in L_2(0, \infty)$ such that

$$[P_T G^{-1}(x_1 - P_T x_1)](t) \neq 0 \tag{VIII. 3-13}$$

on some positive measure subset of $(0, T)$. Since $(x_1 - P_T x_1)(t) = 0$ for $t \leq T$, let $x_1 - P_T x_1 = S_T x_2$ with $x_2 \in L_2(0, \infty)$. Then $G^{-1} S_T x_2 = x_3$ or

$$Gx_3 = S_T x_2 \tag{VIII. 3-14}$$

with $P_T x_3 \neq 0$ [see (VIII. 3-13)]. There is a unique $x_4 \in L_2(0, \infty)$ such that $x_2 = Gx_4$ (since G is a invertible). Then, using (VIII. 3-10),

$$GS_T x_4 = S_T Gx_4 = S_T x_2 \tag{VIII. 3-15}$$

which implies $S_T x_4 = x_3$ [compare (VIII. 3-14) and (VIII. 3-15)]. This is contradictory since $P_T x_3 \neq 0$ and $P_T S_T x_4 = 0$.

Lemma 2

Let $h \in L_1(0, \infty)$ and let L be the linear mapping of $L_2(0, \infty)$ into itself defined by

$$(Lx)(t) = \int_0^t h(t - u)x(u)\, du, \qquad x \in L_2(0, \infty). \tag{VIII. 3-16}$$

Let

$$H(s) = \int_0^\infty h(t)e^{-st}\,dt, \qquad s = \sigma + i\omega, \, \sigma \geq 0. \qquad \text{(VIII. 3-17)}$$

Suppose that for some real c,

$$1 + cH(s) \neq 0 \qquad \text{for} \qquad \sigma \geq 0. \qquad \text{(VIII. 3-18)}$$

Then

(i) $(I + cL)$ is invertible on $L_2(0, \infty)$.

(ii) $\|(I + cL)^{-1}\| \leq \sup_{\omega \in (-\infty, \infty)} |[1 + cH(i\omega)]^{-1}|,$ \qquad (VIII. 3-19)

$\|L(I + cL)^{-1}\| \leq \sup_{\omega \in (-\infty, \infty)} |H(i\omega)[1 + cH(i\omega)]^{-1}|.$

$$\text{(VIII. 3-20)}$$

Proof. Given in Appendix VIII. B.

Lemma 2 shows that $(I + cL)$ is invertible on $L_2(0, \infty)$. We want to show that $(I + cL)$ is invertible on $L_{2e}(0, \infty)$. From Lemma 1, $P_T(I + cL)^{-1}x = P_T (I + cL)^{-1}P_T x$ for all $T \in (0, \infty)$ and all $x \in L_2(0, \infty)$. Define a map of $L_{2e}(0, \infty)$ into itself by

$$x_T = P_T(I + cL)^{-1}y_T, \qquad y \in L_{2e}(0, \infty), \qquad \text{all } T \in (0, \infty). \qquad \text{(VIII. 3-21)}$$

This is easily seen to be the inverse of $(I + cL)$ considered as a map of $L_{2e}(0, \infty)$ into itself [if $x \in L_{2e}(0, \infty)$, then $P_T(I + cL)^{-1} (I + cL)x = x_T$ for all $T \in (0, \infty)$ or, in other words, $(I + cL)^{-1} (I + cL)x(t) = x(t)$ for all $t \geq 0$].

All the conditions of Theorems VIII. 2-1 and VIII. 2-2 are now seen to be satisfied; the condition $\eta_\lambda k_\lambda < 1$ in both cases [using (VIII. 3-5) and (VIII. 3-6) for Theorems VIII. 2.1 and VIII. 2.2, respectively] reduces to (VIII. 3-9).

VIII.4 DISCUSSION

In the last section, the linear operator L mapped $X = L_2(0, \infty)$ into itself although the Theorems of Sec. VIII. 2 only require that L map $X_e = L_{2e}(0, \infty)$ into itself. We imposed the stronger requirement on L in Sec. VIII. 3 only for the sake of simplicity. When we relax the requirement on L we allow, e.g., integrators. Let $y = Lx$ be defined by

$$y(t) = \int_0^t [h_1(t - u) + h_2(t - u)]x(u)\,du, \qquad x \in L_{2e}(0, \infty), \qquad \text{(VIII. 4-1)}$$

where h_1 and h_2 are real-valued measurable functions on $[0, \infty)$ with h_1 bounded and $h_2 \in L_1(0, \infty)$. If $h_1(t) = 1$ and $h_2(t) = 0$ for $t \geq 0$ we have an integrator. This convolution operator does not necessarily map $L_2(0, \infty)$ into itself. For example, let $h_1(t) = 1$ for $t \geq 0$ and $x \in L_2(0, \infty)$, $x \neq 0$, $x(t) \geq 0$ for $t \geq 0$. Then

$$\int_0^t h_1(t - u)x(u)\,du = \int_0^t x(u)\,du \geq m > 0 \qquad \text{(VIII. 4-2)}$$

for all t sufficiently large and thus $\int_0^t x(u)\,du \notin L_2(0, \infty)$.

However, it is easily seen that the convolution of (VIII. 4-1) does map $L_{2e}(0, \infty)$ into itself.

A result for a case of the convolution operator (VIII. 4-1) with $h_1(t) = k_1$, a constant, and $n(x(t), t) = b(t)$ (a measurable function), $\alpha \leq b(t) \leq \beta$, is given in Sandberg [5]. With

$$H(s) = \frac{k_1}{s} + \int_0^\infty h_2(t)e^{-st}\,dt \qquad \text{(VIII. 4-3)}$$

for $s \in S = \{s : s \neq 0, \sigma \geq 0\}$, a stability condition (for the time-varying linear system) is given as

$$1 + \tfrac{1}{2}(\alpha + \beta)H(s) \neq 0 \qquad \text{for all } s \in S \qquad \text{(VIII. 4-4)}$$

$$\sup_{\omega > 0}\left|\frac{H(i\omega)}{1 + \tfrac{1}{2}(\alpha + \beta)H(i\omega)}\right|\tfrac{1}{2}(\beta - \alpha) < 1. \qquad \text{(VIII. 4-5)}$$

We mention this result to point out that the previous geometrical interpretations do not generally apply. For example, let $\alpha = 0$ and $h_2(t) = 0$, all $t \geq 0$. Then

$$\sup_{\omega > 0}\left|\frac{k_1/i\omega}{1 + \beta k_1/2i\omega}\right|\frac{\beta}{2} = 1 \qquad \text{(VIII. 4-6)}$$

so that (VIII. 4-5) is not satisfied. But note that $\operatorname{Re}[H(i\omega)] = 0 > -\beta^{-1}$.

Exercise VIII. 4.1

Why doesn't the geometrical interpretation apply?

In the last section, we interpreted the theorems of Sec. VIII. 2 in $L_2(0, \infty)$ primarily for simplicity. A space which is perhaps more appealing from a practical point of view is

$L_\infty(0, \infty) = \{x : x$ a real valued measurable function defined on $[0, \infty)$,

$$\sup_{t>0}|x(t)| < \infty\} \qquad\qquad \text{(VIII. 4-7)}$$

with norm

$$||x||_\infty = \sup_{t>0}|x(t)|. \qquad\qquad \text{(VIII. 4-8)†}$$

This space is similar to a space of continuous functions we have discussed earlier but is more general. With this space, it is natural to discuss bounded inputs and outputs. Bounded inputs are often of more interest than square integrable inputs. Also, the $L_\infty(0, \infty)$ norm information on the system response is often more valuable than $L_2(0, \infty)$ information (the beginning of Sec. V.1 is somewhat relevant). Furthermore, note that nontrivial periodic functions cannot be in $L_2(0, \infty)$ but can be treated in $L_\infty(0, \infty)$ (only steady state responses were considered in Chapters 4 and 5). With the background obtained from this chapter, the reader can consult the literature for $L_\infty(0, \infty)$ results.

However, we do not wish to dismiss $L_2(0, \infty)$ as only of tutorial value. If we make the assumptions that $h \in L_2(0, \infty)$ and $\lim_{t\to\infty} r(t) = 0$ then $y \in L_2(0, \infty)$ and

$$x(t) = -\int_0^t h(t - \tau)y(\tau)\, d\tau + r(t), \qquad t \geq 0, \qquad \text{(VIII. 4-9)}$$

imply

$$\lim_{t\to\infty} x(t) = 0. \qquad\qquad \text{(VIII. 4-10)}$$

To show this, let

$$p(t) = \int_0^t h(t - u)y(u)\, du = \frac{1}{2\pi}\int_{-\infty}^{\infty} H(i\omega)K(i\omega)e^{i\omega t}\, d\omega. $$

$$\text{(VIII. 4-11)}$$

[Where does (VIII. 4-11) come from?] Since $H(i\omega)K(i\omega)$ is the product of two square integrable functions, it is absolutely integrable [condition (III. 3-2) holds for complex functions]. Hence, $p(t)$ has a representation similar to that of a Fourier transform of an $L_1(\infty, \infty)$ function. The Riemann–Lebesgue lemma is applicable here also to imply

$$\lim_{t\to\infty} p(t) = 0. \qquad\qquad \text{(VIII. 4-12)}$$

Also, $p(t)$ is bounded [see (III. 3-5)] so that $x(t)$ is also bounded if $r(t)$ is. Hence, the $L_2(0, \infty)$ results can, with minor additional assumptions, provide some boundedness and asymptotic information.

† More precisely, the supremum is replaced by the essential supremum.

The desirability of using $L_2(0, \infty)$ arises in part because it is a Hilbert space. In the next section the presence of an inner product will be fully exploited.

VIII.5 USE OF POSITIVITY IN INNER PRODUCT SPACES

We return to the abstract setting of Sec. VIII.2, but now further assume that X is a real† inner product space with inner product $\langle \cdot, \cdot \rangle$. The norm of $x \in X$ is $\|x\| = \langle x, x \rangle^{1/2}$. $L_2(0, \infty)$ is an inner product space but in Sec. VIII.3 we made no use of the inner product since we were applying the results of Sec. VIII. 2 which are for normed linear spaces. By taking advantage of the inner product in $L_2(0, \infty)$, we shall find that conditions for boundedness can be considerably relaxed. The notion of positivity of an operator will be found to be useful in this connection.

If F maps X into itself we say that F is *positive* if for all $x \in X$

$$\langle x, Fx \rangle \geq 0. \qquad \text{(VIII. 5-1)}$$

F is *strongly positive* if there is a a $\delta > 0$ such that for all $x \in X$,

$$\langle x, Fx \rangle \geq \delta \langle x, x \rangle. \qquad \text{(VIII. 5-2)}$$

If F maps X_e into itself, F is *e-positive* if for all $x \in X_e$, all $T \in \Omega$,

$$\langle x_T, (Fx)_T \rangle \geq 0. \qquad \text{(VIII. 5-3)}$$

F is *strongly e-positive* if there is a $\delta > 0$ such that for all $x \in X_e$, all $T \in \Omega$,

$$\langle x_T, (Fx)_T \rangle \geq \delta \langle x_T, x_T \rangle. \qquad \text{(VIII. 5-4)}$$

An example of a positive operator on $L_2(0, \infty)$ is the linear operator L defined by (VIII. 3-1) with $\text{Re}\,[H(i\omega)] \geq 0$ for all real ω since

$$\langle x, Lx \rangle = \int_0^\infty x(t)(Lx)(t)\,dt = \frac{1}{2\pi}\int_{-\infty}^\infty \overline{X(i\omega)}H(i\omega)X(i\omega)\,d\omega$$

$$= \frac{1}{2\pi}\int_{-\infty}^\infty |X(i\omega)|^2 H(i\omega)\,d\omega \geq \inf \text{Re}[H(i\omega)]\,\langle x, x \rangle \qquad \text{(VIII. 5-5)}$$

where $X(i\omega)$ and $H(i\omega)$ are the Fourier transforms of x and h, respectively, and where we have used (III. 3-18) and the evenness and oddness of the real and imaginary parts of the Fourier transform of a real function. L also maps L_{2e}

† Results are not limited to real inner product spaces. We make this assumption only for simplicity of presentation.

$(0, \infty)$ into itself and the condition of e-positivity is demonstrated in a similar manner.

As an example of a nonlinear positive operator on $L_2(0, \infty)$ consider $y = Nx$ defined by

$$y(t) = n(x(t)), \tag{VIII. 5-6}$$

where $n(\sigma)$ is a continuous function of σ for $\sigma \in (-\infty, \infty)$ and

$$0 \leq \frac{n(\sigma)}{\sigma} \leq \beta < \infty, \qquad \text{all real } \sigma \neq 0. \tag{VIII. 5-7}$$

Note that the continuity of n together with (VIII. 5-7) implies $n(0) = 0$. To show positivity,

$$\langle x, Nx \rangle = \int_0^\infty x(t) n(x(t)) \, dt \geq \int_0^\infty 0 \, dt = 0 \tag{VIII. 5-8}$$

using the left-hand inequality of (VIII. 5-7). Note that we did not use the right-hand inequality of (VIII. 5-7) to show positivity ; it was used only to guarantee that $L_2(0, \infty)$ was mapped into itself. N is similarly shown to be strongly positive if it additionally satisfies $n(\sigma)/\sigma \geq \delta > 0$ for all real $\sigma \neq 0$. Also, e-positivity and strong e-positivity are easily demonstrated for $n(\sigma)/\sigma \geq 0$ and $n(\sigma)/\sigma \geq \delta > 0$, respectively.

The next theorem provides a simple sufficient condition for boundedness.

Theorem VIII.5.1

Let L and N map X_e into itself and both be e-positive and suppose either

(a) L is strongly e-positive and $g(L) < \infty$ or

(b) N is strongly e-positive and $g(N) < \infty$.

Let $x \in X_e$, $r \in X$, and

$$x = -LNx + r. \tag{VIII. 5-9}$$

Then $x \in X$ and the system is bounded.

Remark. At this point, we are not assuming L is linear although our main interest will be in this case.

Proof. Denote Nx by z and let $T \in \Omega$. Then,

$$\langle r_T, z_T \rangle = \langle r_T, z_T \rangle - \langle (Lz)_T, z_T \rangle + \langle z_T, (Lz)_T \rangle$$
$$= \langle x_T, (Nx)_T \rangle + \langle z_T, (Lz)_T \rangle \tag{VIII. 5-10}$$

Suppose L is strongly e-positive and $g(L) < \infty$. Then, using the e-positivity of N and the Schwarz inequality with (VIII. 5-10) and also using (VIII. 2-2),

$$\delta\|z_T\|^2 \leq \|r_T\| \cdot \|z_T\|,$$

$$\|z_T\| \leq \delta^{-1}\|r\|. \qquad \text{(VIII. 5-11)}$$

Hence $z \in X$ (see the second paragraph of Sec. VIII. 2). That $x \in X$ follows from (VIII. 5-9),

$$\|x_T\| \leq \|(Lz)_T\| + \|r_T\|$$

$$\leq g(L)\delta^{-1}\|r\| + \|r\|, \qquad \text{all } T \in \Omega. \qquad \text{(VIII. 5-12)}$$

The case where N is strongly e-positive and $g(N) < \infty$ is treated in a similar manner.

Exercise VIII.5.1

Give the proof for N strongly e-positive and $g(N) < \infty$.

Theorem VIII. 5.1 applied to the L and N discussed just prior to the theorem [N described by (VIII. 5-7)] leads to the following sufficient condition for boundedness :

$$n(\sigma)/\sigma \geq \delta > 0 \qquad \text{all real } \sigma \neq 0,$$

$$\text{Re}\,[H(i\omega)] \geq 0, \qquad \text{all } \omega \geq 0. \qquad \text{(VIII. 5-13)}$$

(It is sufficient to consider only nonnegative ω since $\text{Re}\,[H(i\omega)]$ is an even function of ω). To verify that the conditions of the theorem are satisfied, first observe that the strong e-positivity and the e-positivity of N and L, respectively, were shown previous to the theorem. Finally, $g(N) = \beta < \infty$.

Note that we did not use the specific value of β from (VIII. 5-7) in getting a criterion for boundedness. In other words, (VIII. 5-13) is a sufficient condition for boundedness for any value of β. In this light, it is of interest to compare (VIII. 5-13) with condition (VIII. 3-9) for any $\alpha > 0$. Referring to the discussion of the geometric interpretation of (VIII. 3-9), we must have

$$\text{Re}\,[H(i\omega)] > -\beta^{-1}, \qquad \text{all } \omega \geq 0. \qquad \text{(VIII. 5-14)}$$

Now, if (VIII. 5-14) is to be satisfied for all $\beta > 0$, then we must have $\text{Re}\,[H(i\omega)] > 0$ for all $\omega \geq 0$. Hence the above application of Theorem VIII. 5.1 is equivalent to what may be regarded as a limiting case (β arbitrarily large, α arbitrarily small) of a result in Sec. VIII. 3.

Condition (VIII. 5-13) is most restrictive, requiring the phase of the transform to not exceed $90°$. The usefulness of the positive operator approach is considerably enhanced when multiplers (to be discussed below) are used and then decidedly stronger results than those of Sec. VIII. 3 can be obtained for $L_2(0, \infty)$.

The key to the multiplier approach is to represent the linear operator L as a composite of two linear operators, $L = L_2L_1$, and to split off L_2 and associate it with N to form a new nonlinearity. Hence, let L_1, L_2, and N map X_e into itself. Furthermore, let L_2 map X_e into X_s, a subset of X_e, and let there be a linear map L_3 of X_s into X_e such that

$$L_3L_2 f = f \qquad \text{for all } f \in X_e$$
$$L_2L_3 g = g \qquad \text{for all } g \in X_s. \qquad \text{(VIII. 5-15)}$$

Then, with $x \in X_e, r \in X_s$,

$$x = -LNx + r = -L_2L_1Nx + r. \qquad \text{(VIII. 5-16)}$$

Note that (VIII. 5-16) puts x into X_s. With $y = L_3x$ or $x = L_2y$,

$$y = -L_1NL_2y + L_3r. \qquad \text{(VIII. 5-17)}$$

We now have the L_1, NL_2, L_3r of (VIII. 5-17) playing the roles of L, N, r, respectively, of (VIII. 5-16). The manipulations are shown in Figure VIII. 6.

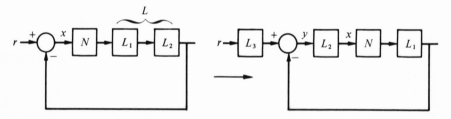

Figure VIII.6 Splitting the linear operator

L_3 is called the multiplier since $L_1 = L_3L$ and conditions on L_1 are derived. This will be clarified below. L_3 will usually be describable in terms of a transfer function.

The following theorem uses a proof similar to that of Theorem VIII. 5.1 but it is more convenient to use with multipliers.

Theorem VIII.5.2

Let L_1 and N map X_e into itself and L_2 map X_e into $X_s \subset X_e$. Assume there is a linear map L_3 of X_s into X_e satisfying (VIII. 5-15). Suppose that for some $\beta \in (0, \infty)$ and some $\delta > 0$

(a) $\langle f_T, (NL_2 f)_T \rangle \geq \beta^{-1} \|(NL_2 f)_T\|^2$

  for all $f \in X_e$, all $T \in \Omega$,

(b) $\langle f_T, (L_1 f)_T \rangle \geq (\delta - \beta^{-1}) \|f_T\|^2$

  for all $f \in X_e$, all $T \in \Omega$,

(c) $g(L) < \infty$.

Let $x \in X_e$, $r \in X \cap X_s$, $L_3 r \in X$, $L = L_2 L_1$, and $x = -LNx + r$. Then $x \in X$.

Remark. Condition (a) implies that NL_2 is e-positive and (b) is equivalent to $(L_1 + \beta^{-1} I)$ being strongly e-positive.

Proof. As in (VIII. 5-17), we have $y = -L_1 NL_2 y + L_3 r$. Denote $NL_2 y$ by z and let $T \in \Omega$. Then, as in (VIII. 5-10),

$$\langle (L_3 r)_T, z_T \rangle = \langle y_T, (NL_2 y)_T \rangle + \langle z_T, (L_1 z)_T \rangle.$$

Using (a) and (b) and the Schwarz inequality,

$$\|(L_3 r)_T\| \cdot \|z_T\| \geq \beta^{-1} \|(NL_2 y)_T\|^2 + (\delta - \beta^{-1}) \|z_T\|^2$$
$$= \delta \|z_T\|^2,$$
$$\|z_T\| \leq \delta^{-1} \|(L_3 r)_T\| \leq \delta^{-1} \|L_3 r\|.$$

Hence $z \in X$. That $x \in X$ follows from

$$\|x_T\| \leq \|(Lz)_T\| + \|r_T\|$$
$$\leq g(L) \delta^{-1} \|L_3 r\| + \|r\|.$$

Remark. The system satisfies a modified form of boundedness; a bound on $\|x\|$ depends on both $\|r\|$ and $\|L_3 r\|$.

Let us apply Theorem VIII.5.2 and let $g = L_2 f$ be described by the following convolution for some $q > 0$ and all $f \in X_e = L_{2e}(0, \infty)$,

$$g(t) = \int_0^t q^{-1} e^{-q^{-1}(t-u)} f(u) \, du, \qquad \text{all } t \geq 0. \qquad \text{(VIII. 5-18)}$$

The $L_1(0, \infty)$ Fourier transform of $q^{-1} e^{-q^{-1} t}$ is $(1 + qi\omega)^{-1}$ so that

$$G_T(i\omega) = (1 + qi\omega)^{-1} F_T(i\omega), \qquad \omega \in (-\infty, \infty), \qquad \text{(VIII. 5-19)}$$

with $G_T(i\omega)$ and $F_T(i\omega)$ the $L_2(0, \infty)$ Fourier transforms of g_T and f_T, respectively. To see what the space X_s is, we recall the equivalence between differentiating a time function and multiplying its Fourier transform by $i\omega$. More precisely, we quote the following two results from Bochner and Chandrasekharan (pp. 124–128).

Theorem VIII.5.3

If f and $\dot{f} \in L_2(0, \infty)$ and the $L_2(0, \infty)$ Fourier transform of f is $F(i\omega)$, then $i\omega F(i\omega)$ is the $L_2(0, \infty)$ Fourier transform of $\dot{f}$.

Theorem VIII.5.4

If $F(i\omega)$ and $i\omega F(i\omega)$ are square integrable over $(-\infty, \infty)$ and $F(i\omega)$ is the Fourier transform of $f \in L_2(0, \infty)$, then $\dot{f}$ exists† and $i\omega F(i\omega)$ is the Fourier transform of $\dot{f}$. Furthermore, if $f, g \in L_2(0, \infty)$, $F(i\omega)$ and $G(i\omega)$ are their respective Fourier transforms, and $G(i\omega) = i\omega F(i\omega)$, then f is differentiable and $g(t) = \dot{f}(t)$.

Using Theorem VIII. 5.4, we see that X_s consists of $L_{2e}(0, \infty)$ functions whose truncations also have derivatives (defined almost everywhere) in $L_2(0, \infty)$. That $\dot{g}_T$ exists and belongs to $L_2(0, \infty)$ follows from (VIII. 5-19) and the fact that $i\omega (1 + qi\omega)^{-1} F_T(i\omega)$ is square integrable over $(-\infty, \infty)$ [since $i\omega (1 + qi\omega)^{-1}$ is bounded] which implies that $i\omega(1 + qi\omega)^{-1} F_T(i\omega)$ is the $L_2(0, \infty)$ Fourier transform of $\dot{g}_T$ (see Theorem VIII. 5.4). A function $g \in X_s$ also satisfies $g(0) = 0$. Furthermore, the linear operator L_3 mapping X_s into X_e is defined by $f = L_3 g$ with

$$F_T(i\omega) = (1 + qi\omega)\, G_T(i\omega), \qquad \omega \in (-\infty, \infty). \qquad \text{(VIII. 5-20)}$$

for all $T > 0$.

Condition (b) of Theorem VIII. 5.2 becomes (see (VIII. 5-5) and using the even and odd properties of Re $[H(i\omega)]$ and Im $[H(i\omega)]$, respectively)

$$\text{Re}\left[(1 + qi\omega)H(i\omega)\right] + \beta^{-1} \geq \delta > 0, \text{ all } \omega \geq 0. \qquad \text{(VIII. 5-21)}$$

Condition (VIII. 5-21) will be referred to as *Popov's condition*. Let us give a geometrical interpretation to (VIII. 5-21); afterwards we will verify that condition (a) of Theorem VIII. 5.2 is satisfied. We rewrite (VIII. 5-21) as

$$\text{Re}\left[H(i\omega)\right] \geq \delta - \beta^{-1} + \omega q \, \text{Im}\left[H(i\omega)\right], \text{ all } \omega \geq 0, \qquad \text{(VIII. 5-22)}$$

which means that *for each ω the locus of $H(i\omega)$ lies to the right of a straight line defined by (VIII. 5-22) with the inequality replaced by equality* (see Figure VIII. 7).

---

† More precisely, f is absolutely continuous and its derivative exists almost everywhere. An introduction to absolutely continuous functions is given in Natanson.

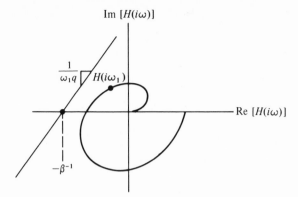

Figure VIII.7 Popov's condition in the $H(i\omega)$ plane

The interpretation of Popov's condition in Figure VIII. 7 is somewhat inconvenient since the slope of the line changes with ω. For this reason, one defines a modified transform $H^*(i\omega)$,

$$H^*(i\omega) = \text{Re}\left[H(i\omega)\right] + i\omega \ \text{Im}\left[H(i\omega)\right]. \qquad \text{(VIII. 5-23)}$$

In terms of $H^*(i\omega)$, Popov's condition reads

$$\text{Re}\left[H^*(i\omega)\right] \geq \delta - \beta^{-1} + q \ \text{Im}\left[H^*(i\omega)\right] \qquad \text{(VIII. 5-24)}$$

which is interpreted in the $H^*(i\omega)$ plane in Figure VIII. 8. The straight line in Figure VIII. 8 is called a *Popov line*. We shall come back to this geometrical interpretation in the next section.

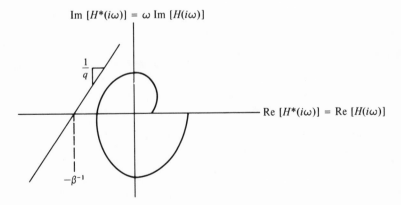

Figure VIII.8 Popov's condition with $H^*(i\omega)$

An unfinished piece of business is to show that condition (a) of Theorem VIII. 5.2 is satisfied. First we show that if $g = L_2 f$ [L_2 described by (VIII. 5-

18)], $f \in L_{2e}(0, \infty)$, $T \in (0, \infty)$, then

$$\langle \dot{g}_T, (Ng)_T \rangle \geq 0, \tag{VIII. 5-25}$$

That $\dot{g}_T \in L_2(0, \infty)$ has already been established. Also, note from (VIII. 5-18) that $g(0) = 0$. Hence,

$$\langle \dot{g}_T, (Ng)_T \rangle = \int_0^T \dot{g}(t)n(g(t)) \, dt = \int_{g(0)}^{g(T)} n(g) \, dg$$

$$= \int_0^{g(T)} n(g) \, dg \geq 0 \tag{VIII. 5-26}$$

the last inequality following from the left-hand inequality of (VIII. 5-7) (i.e., the graph of n lies in the first and third quadrants).

With $f \in L_{2e}(0, \infty)$, $g = L_2 f$,

$$f_T(t) = g_T(t) + q\dot{g}_T(t). \tag{VIII. 5-27}$$

This follows from (VIII. 5-20) and Theorem VIII. 5.4. Then using (VIII. 5-26),

$$\langle f_T, (NL_2 f)_T \rangle = \int_0^T g(t)n(g(t)) \, dt + \int_0^T q\dot{g}(t)n(g(t)) \, dt$$

$$\geq \int_0^T g(t)n(g(t)) \, dt. \tag{VIII. 5-28}$$

Using the following implication of (VIII. 5-7),

$$\beta \sigma n(\sigma) \geq [n(\sigma)]^2, \tag{VIII. 5-29}$$

(VIII. 5-28) becomes

$$\langle f_T, (NL_2 f)_T \rangle \geq \beta^{-1} \int_0^T [n(g(t))]^2 \, dt = \beta^{-1} \|(NL_2 f)_T\|^2. \tag{VIII. 5-30}$$

Recapitulating the key conditions of Theorem VIII. 5.2 for the system described by (VIII. 3-1) and (VIII. 5-6) :

$$0 \leq \frac{n(\sigma)}{\sigma} \leq \beta < \infty, \quad \text{all real } \sigma \neq 0 \, ;$$

$$\text{Re} \left[(1 + qi\omega)H(i\omega) \right] + \beta^{-1} \geq \delta > 0, \quad \text{some } q > 0, \quad \text{all } \omega \geq 0 \, ;$$

$$r, \dot{r} \in L_2(0, \infty), \quad \dot{h} \in L_1(0, \infty). \tag{VIII. 5-31}$$

We added the conditions $\dot{r} \in L_2(0, \infty)$ and $\dot{h} \in L_1(0, \infty)$ for satisfaction of $L_3 r \in L_2(0, \infty)$ and $L_1 = L_3 L$ mapping $L_{2e}(0, \infty)$ into itself, respectively, which are required by Theorem VIII. 5.2. The second condition is justified using a result for $L_1(0, \infty)$ functions in Bochner and Chandrasekharan, p.8, (analogous to Theorem VIII. 5.3) implying that $i\omega H(i\omega)$ is the $L_1(0, \infty)$ Fourier transform of $\dot{h}$. The condition $g(L) < \infty$ is satisfied with $H(i\omega)$ the Fourier transform of an $L_1(0, \infty)$ function.

VIII.6 POPOV'S CONDITION AND AIZERMAN'S CONJECTURE

We know from Sec. VI. 4 that Aizerman's conjecture is not always valid. Popov's condition can delineate classes of systems for which the conjecture is valid.

We consider the same system as used to derive Popov's condition. Recall that

$$(Nx)(t) = n(x(t)), \qquad n(0) = 0, \qquad \frac{n(\sigma)}{\sigma} \in [0, \beta] \qquad \text{for } \sigma \neq 0$$

$$(\text{VIII. 6-1})$$

and $H(i\omega)$ is the Fourier transform of the $L_1(0, \infty)$ impulse response function h. In this section, we make the additional simplifying assumption that $H(i\omega)$ is a rational function of $i\omega$ so that when N is linear, i.e., $n(u) = ku$, the familiar Nyquist's criterion is a necessary and sufficient condition for the poles of the closed loop transfer function to be inside the left half-plane.

Exercise VIII.6.1

Show that the satisfaction of Nyquist's criterion implies $L_2(0, \infty)$ boundedness of the linear feedback system.

Since the rational $H(i\omega)$ is the Fourier transform of a function which is in $L_2(0, \infty)$ [it is in $L_1(0, \infty)$ by assumption] we see that $x \in L_2(0, \infty)$, $r(t) \to 0$ as $t \to \infty \Rightarrow x(t) \to 0$ as $t \to \infty$ for the nonlinear feedback system considered here [see discussion of (VIII. 4-9)]. In particular, if $r = 0$, $L_2(0, \infty)$ boundedness implies $x(t) \to 0$ if $r(t) \to 0$ as $t \to \infty$.

From Sec. VI. 4 we recall that the intersection of the locus of $H(i\omega)$ furthest to the left on the real axis defines a range of linear gains for stability (see

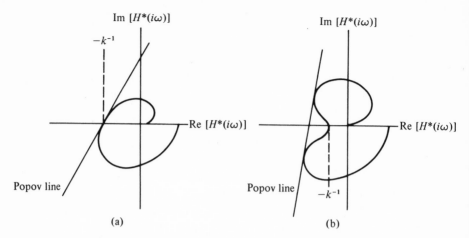

Figure VIII.9 Popov's condition and Aizerman's conjecture

Figure VI. 3). If we now consider the boundedness of systems not necessarily linear and we find the largest k_1 so that the system is bounded if $n(\sigma)/\sigma \in [0, k_1)$ for $\sigma \neq 0$, we must have that $k_1 \leq k$. If $k_1 = k$, then Aizerman's conjecture is verified and if $k_1 < k$, then Aizerman's conjecture is false.

Observing that $H(i\omega)$ and $H^*(i\omega)$ cut the real axis at the same points, we can relate Popov's condition and Aizerman's conjecture. In Figure VIII. 9(a), we see that a Popov line can be drawn through the real axis at any point to the left of($- k^{-1}, 0$). Hence $k_1 = k$ and Aizerman's conjecture is verified. On the other hand in Figure VIII. 9(b), the point on the real axis furthest to the right with a Popov line cutting it is to the left by a positive minimum distance from $(- k^{-1}, 0)$. Hence Aizerman's conjecture is not verified. It is not disproved by this example since Popov's condition is only a sufficient condition. However, the knowledge that a system violating Aizerman's conjecture must also violate a Popov condition is clearly useful in looking for counterexamples to the conjecture.

Let us reconsider the example of the failure of Aizerman's conjecture in Sec. VI. 4. The transfer function and nonlinearity from Sec. VI. 4 are

$$G(i\omega) = \frac{(i\omega)^2}{[(i\omega)^2 + 1][(i\omega)^2 + 9] + \epsilon[\alpha(i\omega)^3 + \beta(i\omega)^2 + \gamma i\omega + \delta]},$$
$$\text{(VIII. 6-2)}$$

$$n(\sigma) = \epsilon\sigma^3 \qquad \text{for} \qquad |\sigma| \leq 1.88. \qquad \text{(VIII. 6-3)}$$

See the comment below (VI. 4-7) and note that σ_m must be at least as large as $\max_t |v^{(2)}(t)|(= \max_t |\sin t - \cos 3t| \simeq 1.88$ with $v(t)$ given by the right-hand side of (VI. 4-19). Hence for $\sigma \neq 0$

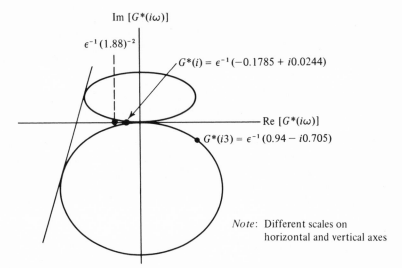

Figure VIII.10 Violation of Popov's condition for counterexample
to Aizerman's conjecture

$$\frac{n(\sigma)}{\sigma} \le \epsilon k_1 \qquad\qquad \text{(VIII. 6-4)}$$

with $k_1 \ge (1.88)^2$. For a given small positive ϵ, the locus of $G^*(i\omega)$ is sketched
in Figure VIII. 10. We do not know how small ϵ must be to be "sufficiently
small" for the weakly nonlinear theory, but we assume that ϵ is small enough.
Then the point $(-k_1^{-1}, 0)$ must certainly lie to the right of the line in Figure
VIII. 10 ; otherwise Popov's condition would be satisfied which would contradict the existence of the periodic solution already established. As $\epsilon \to 0$, $G(i\omega)$
remains the Fourier transform of a function in $L_1(0, \infty) \cap L_2(0, \infty)$ but the loci
of $G(i\omega)$ and $G^*(i\omega)$ grow in size in the complex plane. For sufficiently small
$\epsilon > 0$, $G(i\omega)$ also has no poles in Re $s \ge 0$. When $\epsilon = 0$, $G(i\omega)$ is no longer
the Fourier transform of a function in either $L_1(0, \infty)$ or $L_2(0, \infty)$ and, in fact,
goes off to infinity when $\omega = 1$ or 3. But for all sufficiently small $\epsilon > 0$, the
point $(-k_1^{-1}, 0)$ must lie to the right of the line in Figure VIII. 10.

In the case of Figure VIII. 9(a), Aizerman's conjecture is verified. In Figure VIII. 9(b) and Figure VIII. 10, Aizerman's conjecture is not verified but,
as previously mentioned, not disproved by the nature of the strongest Popov
condition result available. When Popov's condition does not verify Aizerman's
conjecture, two roads are open. One is to prove that the conjecture is indeed
invalid. This is what was done in connection with Figure VIII. 10. Thus Popov's condition can be used to delineate classes of systems for which Aizerman's
conjecture *might* be violated.

On the other hand, Popov's condition assumes rather little information
about the nonlinearity, only that the graph lies in a certain sector. As men-

tioned previously, this is both a strength in not requiring detailed information and a weakness in not taking advantage of some additional information that may be available. If more is known about the nonlinearity (e.g., it is odd or nondecreasing) then it may be possible to sharpen the stability criterion with the potential for verification of Aizerman's conjecture. This is currently a subject of intensive research. The technique of multipliers is widely used in this context, the multipliers being more general than the simple multiplier we used in connection with Popov's condition.† The reader should, at this point, be able to consult the recent literature (e.g., Zames [3]). It should be noted that stability conditions with multipliers can get complicated to apply. Some work in removing multipliers from the final statement of the stability conditions has been done, at least for linear time varying systems (Freedman).

VIII.7   MORE DISCUSSION

We have taken a functional analysis approach towards stability in this chapter. We have not discussed Lyapunov's second method because excellent references on that topic already exist and we wished to present a functional analysis approach. A lucid introduction to Lyapunov's second method is given in the book by LaSalle and Lefschetz. There is a Lyapunov function approach to conditions such as Popov's. The bridge between the Lyapunov functions and frequency domain conditions is crossed with some results due to Kalman and Yacubovich. These matters are discussed in the books by Aizerman and Gantmacher and by Lefschetz (also see K. R. Meyer). We pointed out in Sec. VIII. 6 that multipliers more general than Popov's can be used with more information on the nonlinearities to obtain sharper stability conditions. This may be done using both the functional analysis and Lyapunov function approaches; a representative paper using the latter approach is Narendra and Neuman.

Popov's conditon was originally derived using some rather specialized arguments (see the derivation in Aizerman and Gantmacher). This type of derivation has been extended in Bergen, Iwens, and Rault. Such derivations lend themselves naturally to bounding transient response (see, e.g., Iwens) and to deriving stability conditions applicable to the cases of the nonlinear functions satisfying conditions in restricted sets (see, e. g, the papers by A. U. Meyer and Holtzman [7]). Lyapunov's second method is useful in determing restricted regions of stability. See, for example, the discussion of the extent of asymptotic stability in LaSalle and Lefschetz. This is also an area of current research.

We do not attempt to make any definitive comments as to the overall status of stability of nonlinear feedback systems. (A summary of some of the work on stability done prior to 1966 is given in Brockett.) The field is so ac-

---

† Some clarification of the general relationship between multipliers and Aizerman's conjecture is given in Brockett and Willems.

tive currently that any such comments could only be very tentative, and not based on adequate perspective. However, the approach presented in this chapter is believed to be quite powerful.

VIII.8 NOTES

Theorems VIII. 2.1 and VIII. 2.2 are based on Sandberg [5]. Appendix VIII.A is based on Doležal. The proof of the Paley-Wiener theorem in Appendix VIII.B follows Katznelson, p. 175 (also see Yosida, p. 163). Section VIII.3 follows Sandberg [6]. The argument surrounding (VIII. 4-11) is given in Sandberg [6], p. 1596. Section VIII.5 is an interpretation of Zames [2] and [3] and Sandberg [7]. Also see Zames and Falb. The definitions of boundedness and continuity of Zames [2] are slightly different from those used here [see the discussion of (VIII. 1-16)]. Some other references on stability are given in Sec. VIII. 7. In this chapter we are more interested in explaining the foundations of a functional analysis approach to stability than in giving the latest results available so that many interesting references are not even mentioned.

In Sec. VIII.2 we stated theorems for boundedness and continuity while in Sec. VIII.5, we only gave a boundedness theorem for using multipliers. We could also have given a continuity result which would have the same relationship to Theorem VIII. 5.1 as Theorem VIII. 2.2 has to Theorem VIII. 2.1. Roughly speaking, the use of positivity would be replaced by

$$\langle x_1 - x_2, \, Fx_1 - Fx_2 \rangle \geq 0$$

which is called monotonicity. To our knowledge, multipliers have not yet been successfully used in extending continuity results. However, monotone operators have been used in much recent work on solvability of equations.† Some applications are discussed in Minty; also see Saaty for some discussion and references.

A functional analysis approach to determining conditions for instability is given in Willems [2]. With $x = -LNx + r$ or $(I + LN)x = r$, instability is discussed in terms of the nonexistence of $(I + LN)^{-1}$ as a nonanticipative map. Such existence is related to the well-posedness or physical realizability of the system described by the feedback equation. Early work on realizability conditions for feedback systems is reported in Zames [4].

---

† Also, monotonicity of the nonlinear function in a feedback loop has been used to derive boundedness conditions (see Zames [3]).

APPENDIX VIII.A

Existence in X_e

In Sec. VIII.2, we assumed existence of solutions to $x = - LNx + r$ in X_e and gave conditions for $x \in X$. We now show how some minor additional assumptions in Theorem VIII. 2.2 can imply $x \in X_e$ so that such existence need not be assumed at the outset.

Let $A \in \Theta$ and define for every $T \in \Omega$

$$X_T = \{x : x = P_T y, \ y \in X_e\}. \tag{VIII. A-1}$$

We shall reduce the problem of finding fixed points of A in X_e to the easier job of finding fixed points of $P_T A$ in the smaller set $X_T \subset X \subset X_e$.

Lemma 1

If for every $T \in \Omega$ the operator $P_T A$ has a unique fixed point in X_T, i.e., a unique $x^{(T)} \in X_T$ exists such that

$$x^{(T)} = P_T A x^{(T)}, \tag{VIII. A-2}$$

then there exists a unique $x \in X_e$ satisfying

$$x = Ax. \tag{VIII. A-3}$$

Proof. Let $T, T' \in \Omega$ and $T \leq T'$; we show that $x_T^{(T')} = x^{(T)}$. From (VIII. A-2)

$$x^{(T)} = P_T A x^{(T)}, \qquad x^{(T')} = P_{T'} A x^{(T')}. \tag{VIII. A-4}$$

Applying P_T to the second equation of (VIII. A-4) and using (VIII. 2-4) and the nonanticipativeness of A, we get

$$x_T^{(T')} = P_T A x^{(T')} = P_T A P_T x^{(T')} = P_T A x_T^{(T')} \tag{VIII. A-5}$$

Since $x_T^{(T')} \in X_T$, it follows from the assumed uniqueness that $x_T^{(T')} = x^{(T)}$.

Next, define $x \in X_e$ as follows: if $t \in \Omega$, put $x(t) = x^{(T)}(t)$ where $T \in \Omega$ and $T \geq t$. To show that this definition is meaningful, let $T' \in \Omega$, $T' \geq T$. We have by the above, $x^{(T)} = x_T^{(T')}$, i.e., $x_T^{(T)} = x_T^{(T')}$ so that $x^{(T)}(\tau) = x^{(T')}(\tau)$ for $\tau \in [T]$ by (VIII. 2-3). Furthermore, for any $T \in \Omega$, we have $x(t) = x^{(T)}(t)$ on $[T]$, i.e., by (VIII. 2-3), $x_T = x_T^{(T)}$. Since $x_T^{(T)} \in X_T \subset X_e$ for all $T \in \Omega$, it follows that $x \in X_e$.

To show that this x satisfies (VIII. A-3), let $T \in \Omega$. Then $x_T = x_T^{(T)} = x^{(T)}$. Moreover, $(Ax)_T = (Ax_T)_T = (Ax^{(T)})_T = x_T$. Hence, from (VIII. 2-3), $(Ax)(\tau) = x(\tau)$ for $\tau \in [T]$ for all $T \in \Omega$ and thus $(Ax)(t) = x(t)$ for all $t \in \Omega$.

To prove uniqueness, assume there is a $y \in X_e$ such that $y = Ay$. Letting $T \in \Omega$, we have $y_T \in X_T$ and $y_T = (Ay)_T = (Ay_T)_T$. From the uniqueness of $x^{(T)}$, $y_T = x^{(T)}$, and, as above, $y_T = x_T$. From (VIII. 2-3), $y(t) = x(t)$ on $[T]$ and it follows that $y(t) = x(t)$ for all $t \in \Omega$.

The easiest way to use Lemma 1 is to show that $P_T A$ is a contraction mapping on X_T and use the CMT. This would require that X_T be a complete space, an assumption not made in Sec. VIII.2. Recall that X was assumed to be a normed linear space but not necessarily complete. The next lemma proves that if X is a Banach space, then so is X_T.

Lemma 2

If X is a Banach space, then so is X_T (with the norm of X) for any $T \in \Omega$.

Proof. X_T is clearly a normed linear space. We must show it is complete, i.e., every Cauchy sequence $\{x_i\}$, $x_i \in X_T$, $i = 1, 2, \ldots$, converges to an $x \in X_T$. Since also $x_i \in X$ and X is complete there is an $x \in X$ such that $x_i \to x$ in X, i.e., $\|x_i - x\| \to 0$ as $i \to \infty$. Using (VIII. 2-2), we have $\|x_i - x_T\| = \|(x_i)_T - x_T\| = \|(x_i - x)_T\| \leq \|x_i - x\| \to 0$. Hence $x_i \to x_T \in X$ and $x = x_T$. Here we have used Exercise I. 2.1(b).

Returning to Sec. VIII.2 we see that the conditions of Theorem VIII. 2.2 (without assuming x_1, $x_2 \in X_e$) are sufficient to imply the existence of a unique $x \in X_e$ satisfying $x = -LNx + r$ if it is also assumed that X is complete and $(N - \lambda I)$ is nonanticipative.

Proof of Lemma 2 in Section VIII.3

The proof of Lemma 2 uses Lemma 3 (due to Paley and Wiener[†]). Let the so-called Hardy-Lebesgue class $H^2(0)$ be the set of all functions of a complex variable $s = \sigma + i\omega$, $F(s)$, that are defined and analytic in the right half-plane $\sigma > 0$ and for each fixed $\sigma > 0$, $F(\sigma + i\omega)$ as a function of ω is square integrable over $(-\infty, \infty)$ and, with a constant not depending on σ,

$$\int_{-\infty}^{\infty} |F(\sigma + i\omega)|^2 \, d\omega \leq \text{constant} < \infty \qquad \text{for all } \sigma > 0. \qquad \text{(VIII. B-1)}$$

Lemma 3

Let $f \in L_2(-\infty, \infty)$ and $f(t) = 0$ for $t < 0$ or $f \in L_2(0, \infty)$. Then

$$F(\sigma + i\omega) = \int_0^{\infty} f(t)e^{-\sigma t} e^{-i\omega t} \, dt, \qquad \sigma > 0, \qquad \text{(VIII. B-2)}$$

† There are several other theorems called Paley–Wiener theorems (see, e.g., Zadeh and Desoer, Sec. 9.5 and Papoulis, pp. 215–217). These referenced theorems are concerned with conditions for $|H(i\omega)|$ to be the *magnitude* of a Fourier transform of a function vanishing for negative t.

belongs to $H^2(0)$. Conversely, let $F(\sigma + i\omega) \in H^2(0)$. Then $F(i\omega)$ exists in the sense of the following limit,

$$\lim_{\sigma \to 0+} \int_{-\infty}^{\infty} |F(i\omega) - F(\sigma + i\omega)|^2 \, d\omega = 0 \qquad \text{(VIII. B-3)}$$

in such a way that inverse Fourier transform

$$f(t) = \frac{1}{2\pi} \operatorname*{l.i.m.}_{N \to \infty} \int_{-N}^{N} F(i\omega) e^{i\omega t} \, d\omega \qquad \text{(VIII. B-4)}$$

vanishes for $t < 0$ and $F(s)$ may be obtained as the one-sided Laplace transform of $f(t)$ (for $\sigma > 0$).

Remarks

   (i) In (VIII. B-4), f is real if $F(-i\omega) = \overline{F(i\omega)}$, all ω. In (VIII. B-3), $F(i\omega)$ is the $L_2(-\infty, \infty)$ Fourier transform of f; this follows from (VIII. B-4) and the uniqueness of $L_2(-\infty, \infty)$ Fourier transforms and inverses.

  (ii) In the proof of Lemma 2 we shall identify functions f in $L_2(-\infty, \infty)$ satisfying $f(t) = 0$ for $t < 0$ with functions f in $L_2(0, \infty)$. Lemma 3 shows how to represent those functions in $L_2(-\infty, \infty)$ which vanish for negative t.

Proof. Let $f \in L_2(0, \infty)$ and $\sigma > 0$. Then $f(t)e^{-\sigma t} \in L_1(0, \infty)$ by the Schwarz inequality [it is clearly in $L_2(0, \infty)$]. Let $F(\sigma + i\omega)$ be the Fourier transform of $f(t)e^{-\sigma t}$,

$$F(\sigma + i\omega) = \int_0^\infty f(t)e^{-\sigma t}e^{i\omega t} \, dt. \qquad \text{(VIII.B-5)}$$

We have from (III. 3-19) that for all $\sigma > 0$,

$$\frac{1}{2\pi} \int_{-\infty}^{\infty} |F(\sigma + i\omega)|^2 d\omega = \int_0^\infty |f(t)|^2 \, e^{-2\sigma t} \, dt \leq \int_0^\infty |f(t)|^2 \, dt < \infty. \qquad \text{(VIII. B-6)}$$

The function $F(s)$ may be seen to be analytic in the open right half-plane Re $s = \sigma > 0$ by differentiating the right-hand side of (VIII. B-5) with respect to $s = \sigma + i\omega$ under the integral sign observing that $f(t)te^{-st}$ as a function of t is absolutely integrable over $[0, \infty)$ for $\sigma > 0$ and recalling that a differentiable function of a complex variable is analytic (see, e.g., Kaplan, p. 116 or Evgrafov, p.34). That is, for $\sigma > 0$,

$$\frac{d}{ds}F(s) = \int_0^\infty (-t)f(t)e^{-st} \, dt. \qquad \text{(VIII. B-7)}$$

To prove the converse, let $G(i\omega)$ be the transform of $g(t) = f(t)e^{-t}$. Hence,

$$G(i\omega) = F(1 + i\omega). \tag{VIII. B-8}$$

Then, $F(\sigma + 1 + i\omega)$ is the transform of $g(t)e^{-\sigma t}$. From (III. 3-19) and (VIII. B-1),

$$\int_{-\infty}^{\infty} |g(t)|^2 e^{-2\sigma t}\, dt = \frac{1}{2\pi} \int_{-\infty}^{\infty} |F(\sigma + 1 + i\omega)|^2\, d\omega \le \text{constant, all } \sigma > -1. \tag{VIII. B-9}$$

For any $a > 0$,

$$\int_{-\infty}^{0} |g(t)|^2 e^{-2\sigma t}\, dt = \int_{-\infty}^{-a} + \int_{-a}^{0} |g(t)|^2 e^{-2\sigma t}\, dt \ge e^{2\sigma a} \int_{-\infty}^{-a} |g(t)|^2\, dt$$

$$+ \int_{-a}^{0} |g(t)|^2 e^{-2\sigma t}\, dt. \tag{VIII. B-10}$$

Since (VIII. B-10) holds for σ arbitrarily large we have from (VIII. B-9) that the first integral on the right-hand side of (VIII. B-10) must be zero or $g(t) = 0$ (a.e.) for $t \in (-\infty, -a]$. Since a may be any positive number, $g(t) = 0$ for $t \in (-\infty, 0)$. This implies that $f(t) = g(t)e^t = 0$ for $t < 0$.

Proof of Lemma 2. Consider first the invertibility of $(I_1 + cL_1)$ defined on $L_2(-\infty, \infty)$ by

$$g(t) = (I_1 + cL_1)f(t) = f(t) + c\int_{-\infty}^{t} h(t - u)f(u)\, du, \qquad f \in L_2(-\infty, \infty). \tag{VIII. B-11}$$

Arguments of Sec. III.4 show that $cH(i\omega) \ne -1$ is equivalent to

$$\inf_{\omega \in (-\infty, \infty)} |1 + cH(i\omega)| > 0 \tag{VIII. B-12}$$

[see (III. 4-15)]. Hence

$$\sup_{\omega \in (-\infty, \infty)} |(1 + cH(i\omega))^{-1}| < \infty. \tag{VIII. B-13}$$

Let $F(i\omega)$ and $G(i\omega)$ be the $L_2(-\infty, \infty)$ Fourier transforms of f and g, respectively. Using the same arguments as given with (III. 4-17) through (III. 4-19) we see that $(I_1 + cL_1)$ has a bounded linear inverse on $L_2(-\infty, \infty)$ with

$$\left\| (I_1 + cL_1)^{-1} \right\|_{L_2(-\infty, \infty)} \le \sup_{\omega \in (-\infty, \infty)} |(1 + cH(i\omega))^{-1}|. \tag{VIII. B-14}$$

Now we wish to show that $(I_1 + cL_1)^{-1}$ maps

$$L_2^+(-\infty, \infty) = \{g : g \in L_2(-\infty, \infty), \; g(t) = 0 \text{ for } t < 0\}$$

$$\text{(VIII. B-15)}$$

into itself. From Lemma 3, if $g \in L_2^+(-\infty, \infty)$, then the Fourier transform of $g(t)e^{-\sigma t}$ for $\sigma > 0$, $G(\sigma + i\omega)$, belongs to $H^2(0)$. With $f = (I_1 + cL_1)^{-1}g$, we have

$$F(i\omega) = (1 + cH(i\omega))^{-1}G(i\omega) \qquad \text{(VIII. B-16)}$$

and we wish to show $F(\sigma + i\omega) \in H^2(0)$ so that $F(i\omega)$ is the $L_2(-\infty, \infty)$ Fourier transform of an $f \in L_2^+(-\infty, \infty)$. Identifying functions in $L_2^+(-\infty, \infty)$ with functions in $L_2(0, \infty)$ we will have proved that $(I + cL)^{-1}$ maps $L_2(0, \infty)$ into itself.

Note that $F(\sigma + i\omega) = (1 + cH(\sigma + i\omega))^{-1}G(\sigma + i\omega)$ will be in $H^2(0)$ if $(1 + cH(s))^{-1}$ is analytic and bounded for Re $s = \sigma > 0$. This is so because the product of the two functions $(1 + cH(s))^{-1}$ and $G(s)$ (both analytic for $\sigma > 0$) is analytic for $\sigma > 0$. Furthermore,

$$\int_{-\infty}^{\infty} |(1 + cH(\sigma + i\omega))^{-1}G(\sigma + i\omega)|^2 \, d\omega$$

$$\leq \sup_{\sigma > 0} |(1 + cH(\sigma + i\omega))^{-1}|^2 \int_{-\infty}^{\infty} |G(\sigma + i\omega)|^2 \, d\omega, \quad \sigma > 0.$$

$$\text{(VIII. B-17)}$$

Since the reciprocal of an analytic function is analytic if the reciprocal does not vanish, (VIII. 3-18) implies that $(1 + cH(s))^{-1}$ is analytic for $\sigma > 0$.

To show that $(1 + cH(s))^{-1}$ is bounded for $\sigma > 0$ we use a theorem of Phragmén and Lindelöf (see Titchmarch [2], p. 177). A slight restatement of this theorem implies that if $B(s)$ is a function analytic in the open right half-plane and continuous up to the imaginary axis and satisfies the following two properties:

(i) $|B(i\omega)| \leq M < \infty, \; \omega \in (-\infty, \infty),$ \hfill (VIII. B-18)

(ii) $|B(s)| \leq k_1 e^{|s|^{k_2}}$ for some $k_1 < \infty$, some $k_2 < 1$, and for all $|s|$ sufficiently large with $\sigma > 0,$ \hfill (VIII. B-19)

then

$$|B(s)| \leq M, \qquad \sigma \geq 0. \qquad \text{(VIII. B-20)}$$

From (VIII. B-13) we see that $(1 + cH(s))^{-1}$ satisfies (i). To show that it satisfies (ii), we observe that $H(s) \to 0$ uniformly as $s \to \infty$ with $\sigma > 0$. This is a property of Laplace transforms of $L_1(0, \infty)$ functions [see the proof on p. 49 of Doetsch which holds for $L_1(0, \infty)$ functions].

Condition (VIII. 3-20) follows from familiar arguments.

References

(This is not an exhaustive bibliography; only those references specifically cited are listed. Many of the listed references contain further lists of references.)

M. A. Aizerman and F. R. Gantmacher, "Absolute Stability of Regulator Systems," (translation from Russian), Holden-Day, Inc., San Francisco, 1964.

P. M. Anselone (editor), *Nonlinear Integral Equations*, The University of Wisconsin Press, Madison, Wisconsin, 1964.

V. E. Beneš, "A Nonlinear Integral Equation in the Marcinkiewicz Space M_2," *Journal of Mathematics and Physics*, Vol. 44, No. 1, pp. 24–35, March 1965.

A. R. Bergen and R. L. Franks, "On Periodic Responses of Nonlinear Feedback Systems," *IEEE Transactions on Automatic Control*, Vol. AC-15, No. 1, February 1970.

A. R. Bergen, R. P. Iwens, and A. J. Rault, "On Input-Output Stability of Nonlinear Feedback Systems," *IEEE Transactions on Automatic Control*, Vol. AC-11, No. 4, pp. 742–744, October 1966.

T. A. Bickart, "The Exponential Describing Function in the Analysis of Nonlinear Systems," *IEEE Transactions on Automatic Control*, Vol. AC-11, No. 3, pp. 491–497, July 1966.

S. Bochner and K. Chandrasekharan, *Fourier Transforms*, Princeton University Press, Princeton, New Jersey, 1949.

R. W. Brockett, "The Status of Stability Theory for Deterministic Systems," *IEEE Transactions on Automatic Control*, Vol. AC-11, No. 3, pp. 596-606, July 1966.

R. W. Brockett and J. L. Willems, "Frequency Domain Stability Criteria, Part I," *IEEE Transactions on Automatic Control*, Vol. AC-10, No. 3, pp. 255-261, July 1965.

V. W. Bryant, "A Remark on a Fixed-Point Theorem for Iterated Mappings," *American Mathematical Monthly*, Vol. 75, No. 4, pp. 399-400, April 1968.

E. W. Cheney, *Introduction to Approximation Theory*, McGraw-Hill Book Company, Inc., New York, 1966.

S. C. Chu and J. B. Diaz, "On 'In the Large' Application of the Contraction Principle," in *Differential Equations and Dynamical Systems*, edited by J. K. Hale and J. P. LaSalle, Academic Press, Inc., New York, 1967, pp. 235-238.

L. Collatz, *Functional Analysis and Numerical Mathematics* (translation from German), Academic Press, Inc., New York, 1966.

C. A. Desoer and M. Y. Wu, "Stability of Multi-loop Linear Time-Invariant Systems," *Journal of Mathematical Analysis and Application*, Vol. 23, pp. 121-129, 1968.

J. Dieudonné, *Foundations of Modern Analysis*, Academic Press, Inc., New York, 1960.

G. Doetsch, *Theorie und Anwendung der Laplace-Transformation*, Dover, New York, 1943.

V. Doležal, "On General Nonlinear and Quasilinear Unanticipative Feedback Systems," *Aplikace Matematiky*, No. 3, 1969.

R. E. Edwards, *Functional Analysis*, Holt, Rinehart and Winston, New York, 1965.

H. H. Ehrmann, "On Implicit Function Theorems and the Existence of Solutions of Nonlinear Equations," *Enseignement Mathematique*, Vol. 9, No. 3, pp. 129-176, 1963.

M. A. Evgrafov, *Analytic Functions* (translation), W. B. Saunders Company, Philadelphia, 1966.

W. H. Fleming, "Functions of Several Variables," Addison-Wesley Publishing Company, Inc., Reading, Massachusetts, 1965.

M. I. Freedman, "L_2-Stability of Time-Varying Systems–Construction of Multipliers wth Prescribed Phase Characteristics," *SIAM Journal on Control*, Vol. 6, No. 4, pp. 559-578, November 1968.

A. Fukuma and M. Matsubara, "Jump Resonance Criteria of Nonlinear Control Systems," *IEEE Transactions on Automatic Control*, Vol. AC-11, No. 4, pp. 699-706, October 1966.

A. Gelb and W. E. VanderVelde, *Multiple-Input Describing Functions and Nonlinear System Design*, McGraw-Hill Book Company, Inc., New York, 1968.

R. R. Goldberg, *Fourier Transforms*, Cambridge University Press, Cambridge, Massachusetts, 1965.

D. Graham and D. McRuer, *Analysis of Nonlinear Control Systems*, John Wiley & Sons, Inc., New York, 1961.

A. Halanay, *Differential Equations: Stability, Oscillations, Time Lags* (translation), Academic Press, Inc., New York, 1966.

J. K. Hale, *Oscillations in Nonlinear Systems*, McGraw-Hill Book Company, Inc., New York, 1963.

H. Hatanaka, "The Frequency Responses and Jump-Resonance Phenomena of Nonlinear Feedback Control Systems," *Transactions of the A.S.M.E.*, *Journal of Basic Engineering*, Ser. D., Vol. 85, pp. 236-242, June 1963.

C. Hayashi, *Nonlinear Oscillations in Physical Systems*, McGraw-Hill Book Company, Inc., New York, 1964.

H. Hochstadt, *Differential Equations*, Holt, Rinehart and Winston, New York, 1964.

J. M. Holtzman [1] "The Use of the Contraction Mapping Theorem with Derivatives in a Banach Space," *Quarterly of Applied Mathematics*, Vol. XXVI, No. 3, pp. 462-465, October 1968. [2] "Nonlinear Distortion in Feedback Systems," *Bell System Technical Journal*, Vol. 47, No. 4, pp. 503-509, April 1968. [3] "Analysis of Statistical Linearization of Nonlinear Control Systems," *SIAM Journal on Control*, Vol. 6, No. 2, pp. 235-243, May 1968. [4] "Contraction Maps and Equivalent Linearization," *Bell System Technical Journal*, Vol. 46, No. 10, pp. 2405-2435, December 1967. [5] "Explicit ε and δ for the Implicit Function Theorem," *SIAM Review*, Vol. 12, No. 2, April 1970. [6] "Sensitivity Analysis and Implicit Functions," *Proceedings of the 6th Annual Allerton Conference on Circuits and Systems Theory*, Univ. of Illinois, pp. 589-597, 1968. [7] "A Local Bounded-Input Bounded-Output Condition for Nonlinear Feedback Systems," *IEEE Transactions on Automatic Control*, Vol. AC-13, No. 5, pp. 585-587, October 1968.

J. C. Hsu and A. U. Meyer, "Modern Control Principles and Applications," McGraw-Hill Book Company, Inc., New York, 1968.

R. Iwens, "Bounds on the Responses of Nonlinear Control Systems," *Journal of the Franklin Institute*, Vol. 285, No. 4, pp. 261-274, April 1968.

L. V. Kantorovich and G. P. Akilov, *Functional Analysis in Normed Spaces* (translation from Russian), The Macmillan Company, New York, 1964.

W. Kaplan, *Operational Methods for Linear Systems*, Addison-Wesley Publishing Company, Reading, Massachusetts, 1962.

Y. Katznelson, *Introduction to Harmonic Analysis*, John Wiley & Sons, Inc., New York, 1968.

A. N. Kolmogorov and S. V. Fomin, *Elements of the Theory of Functions and Functional Analysis*, Vol. 1, *Metric and Normed Spaces* (translation from Russian), Graylock Press, Rochester, New York, 1957.

M. A. Krasnosel'skii,[†] *Topological Methods in the Theory of Nonlinear Integral Equations* (translation), Pergamon Press Limited, Oxford, England, 1963.

M. A. Krasnosel'kiy[†] et al., *Plane Vector Fields* (translation), Academic Press, Inc., New York, 1966.

[†]Different translations of the same name.

J. Kudrewicz, "Theorems on the Existence of Periodic Vibrations Based upon the Describing Function Method," Fourth Congress of the International Federation of Automatic Control, Warsaw, June 1969.

S. H. Kyong. "Jump Criteria of Nonlinear Control Systems and the Validity of Statistical Linearization Approximation," *Bell System Technical Journal*, Vol. 48, No. 7, pp. 2529-2543, September 1969.

J. P. LaSalle and S. Lefshetz, *Stability by Liapunov's Direct Method with Applications*, Academic Press, Inc., New York, 1961.

S. Lefschetz, *Stability of Nonlinear Control Systems*, Academic Press, Inc., New York, 1965.

G. G. Lorentz, *Approximation of Functions*, Holt, Rinehart and Winston, New York, 1966.

N. W. McLachlan, *Ordinary Differential Equations in Engineering and Physical Sciences*, 2nd Edition, Oxford at the Clarendon Press, 1956.

A. U. Meyer, "Computation of Initial State Regions for System Stability via Frequency Response Criterion," Fourth Congress of the International Federation of Automatic Control, Warsaw, June 1969.

K. R. Meyer, "On the Existence of Lyapunov Functions for the Problem of Lur'e," *SIAM Journal on Control*, Vol. 3, No. 3, pp. 373-383, 1965.

N. Minorsky, *Nonlinear Oscillations*, D. Van Nostrand Company, Inc., Princeton, New Jersey, 1962.

G. J. Minty, "Monotone Operators and Certain Systems of Nonlinear Ordinary Differential Equations," *Proceedings of Symposium on System Theory*, Polytechnic Institute of Brooklyn, pp. 39-55, 1965.

K. S. Narendra and C. P. Neuman, "Stability of a Class of Differential Equations with a Single Monotone Nonlinearity," *SIAM Journal on Control*, Vol. 4, No. 2, pp. 295-308, May 1966.

M. Z. Nashed, "Some Remarks on Variations and Differentials," *American Mathematical Monthly*, Vol. 73, No. 4, Part II, pp. 63-76, April 1966.

I. P. Natanson, *Theory of Functions of a Real Variable*, Vol. I (translation from Russian), Frederick Ungar Publishing Company, New York, 1961.

A. Papoulis, *The Fourier Integral and Its Applications*, McGraw-Hill Book Company, Inc., New York, 1962.

H. R. Pitt, *Tauberian Theorems*, Oxford University Press, New York, 1958.

W. A. Porter, *Modern Foundations on Systems Engineering*, The Macmillan Company, New York, 1966.

T. L. Saaty, *Modern Nonlinear Equations*, McGraw-Hill Book Company, Inc., New York, 1967.

I. W. Sandberg [1] "A Note on the Application of the Contraction-Mapping Fixed-Point Theorem to a Class of Nonlinear Functional Equations," *SIAM Review*, Vol. 7, No. 2, pp. 199-204, April 1965. [2] "Signal Distortion in Nonlinear Feedback Systems," *Bell System Technical Journal*, Vol. XLII, No. 6, pp.

2533–2550, November 1963. [3] "On the Response of Nonlinear Control Systems to Periodic Input Signals," *Bell System Technical Journal*, Vol. XLIII, No. 3. pp. 911–926, May 1964. [4] "On Truncation Techniques in the Approximate Analysis of Periodically Time-Varying Nonlinear Networks," *IEEE Transactions on Circuit Theory*, Vol. CT-11, No. 2, pp. 195–201, June 1964. [5] "An Observation Concerning the Application of the Contraction-Mapping Fixed-Point Theorem, and a Result Concerning the Norm-Boundedness of Solutions of Nonlinear Functional Equations," *Bell System Technical Journal*, Vol. XLIV, No. 8, pp. 1809–1812, October 1965. [6] "On the L_2-Boundedness of Solutions of Nonlinear Functional Equations," *Bell System Technical Journal*, Vol. XLIII, No. 4, Part 2, pp. 1581–1599, July 1964. [7] "Some Stability Results Related to those of V. M. Popov," *Bell System Technical Journal*, Vol. XLIV, No. 9, pp. 2133–2148, November 1965.

T. E. Stern, *Theory of Nonlinear Networks and Systems*, Addison-Wesley Publishing Company, Reading, Massachusetts, 1965.

J. J. Stoker, *Nonlinear Vibrations*, Interscience Publishers, Inc., New York, 1950.

R. A. Struble, *Nonlinear Differential Equations*, McGraw-Hill Book Company, Inc., New York, 1962.

A. E. Taylor, *Advanced Calculus*, Ginn and Company, Boston, 1955.

E. C. Titchmarsh [1] *Introduction to the Theory of Fourier Integrals*, 2nd Edition, Oxford at the Clarendon Press, 1948. [2] *The Theory of Functions*, 2nd Edition, Oxford University Press, London, 1964.

G. P. Tolstov, *Fourier Series* (translation from Russian), Prentice-Hall, Inc., Englewood Cliffs, New Jersey, 1962.

F. G. Tricomi, *Integral Equations*, Interscience Publishers, Inc., New York, 1957.

M. Urabe and A. Reiter, "Numerical Computation of Nonlinear Forced Oscillations by Galerkin's Procedure," *Journal of Mathematical Analysis and Applications*, Vol. 14, pp. 107–140, 1966.

J. C. Willems [1] "Perturbation Theory for the Analysis of Instability in Nonlinear Feedback Systems," *Proceedings of the 1966 Allerton Conference on Circuit and System Theory*. [2] "Stability, Instability, Invertibility and Causality, "*SIAM Journal on Control*, Vol. 7, No. 4, pp. 645–671, November 1969.

K. Yosida, *Functional Analysis*, 2nd Edition, Springer-Verlag, New York, 1968.

L. A. Zadeh and C. A. Desoer, *Linear System Theory*, McGraw-Hill Book Company, Inc., New York, 1963.

G. Zames [1] "Functional Analysis Applied to Nonlinear Feedback Systems," *IEEE Transactions on Circuit Theory*, Vol. CT-10, pp. 392–404, September 1963. [2] "On the Input-Output Stability of Time-Varying Nonlinear Feedback Systems, Part I: Conditions Derived Using Concepts of Loop Gain, Conicity, and Positivity," *IEEE Transactions on Automatic Control*, Vol. AC-11, No. 2, pp. 228–238, April 1966. [3] "On the Input-Output Stability of Time-Varying Nonlinear Feedback Systems–Part II: Conditions Involving Circles in the Frequency Plane and Sector Nonlinearities," *IEEE Transactions on Automatic Control*, Vol.

AC-11, No. 3, pp. 465-476, July 1966. [4] "Realizability Conditions for Nonlinear Feedback Systems," *IEEE Tranactions on Circuit Theory*, Vol. CT-11, pp. 186-194, June 1964.

G. Zames and P. L. Falb, "Stability Conditions for Systems with Monotone and Slope-Restricted Nonlinearities," *SIAM Journal on Control*, Vol. 6, No. 1, pp. 89-108, February 1968.

Name Index

Subject Index

Aizerman's conjecture, 137, 191 ff.
Almost everywhere, 51
Almost periodic function, 74
Approximation theory, 119
Approximations, 37
Argument, principle of, 114, 177
Asymptotically stable, 166
Autonomous system, 110 ff., 131 ff.

Ball:
  closed, 8, 9
  open, 8
Bifurcation equation, 127
Bounded variation, function of, 61, 170
Boundedness:
  of a feedback system, 169, 172
  of a linear operator, 13

Cauchy sequence, 10
Chain rule, 18
Circle condition, 55, 178
Closed, 9
Closure, 9
Complement, 6
Completeness:
  of a metric space, 10
  of a system, 69
Composition, 13

Connected set (in complex plane), 177
Continuity:
  of a feedback system, 169, 172
  of a mapping, 13
Contraction, 24
  global, 33
  local, 33
Contraction constant, 24
Contraction mapping theorem, 24
  interpretation, 38
Convergence, 8
  uniform, 20
Convex, 11
Convolution, 52
Countable, 7
Critical case, 96, 126

Deadzone, 55
Denumerable, 7
Derivative:
  Fréchet, 18
  Gateaux, 17
  partial, 146, 160
Describing function, 79
  dual input, 81, 86, 139
  failure of, 138
  realness, 80
Determining equation, 127